The Strands of a Life

The Strands of a Life

The Science of DNA and the Art of Education

Robert L. Sinsheimer

UNIVERSITY OF CALIFORNIA PRESS

Berkeley / Los Angeles / London

University of California Press
Berkeley and Los Angeles, California

University of California Press, Ltd.
London, England

© 1994 by
The Regents of the University of California

Library of Congress Cataloging-in-Publication Data
Sinsheimer, Robert.
  The strands of a life: the science of DNA and the art of
    education. / Robert L. Sinsheimer.
  p. cm.
  Includes bibliographical references (p.) and index.
  ISBN 0-520-08248-6 (alk. paper)
  1. Sinsheimer, Robert. 2. Molecular biologists—United States—
Biography. 3. College administrators—United States—Biography.
I. Title.
QH31.S573A3 1994   93-13122
574.8'8'092—dc20   CIP

Printed in the United States of America
9 8 7 6 5 4 3 2 1

To Karen

Contents

Illustrations follow p. 156

Foreword

This book was commissioned as part of the Alfred P. Sloan Foundation Science Book Series. The Alfred P. Sloan Foundation has for many years had an interest in encouraging public understanding of science. Science in this century has become a complex endeavor. Scientific statements may reflect many centuries of experimentation and theory and are likely to be expressed in the language of advanced mathematics or in highly technical terms. As scientific knowledge expands, the goal of general public understanding of science becomes increasingly difficult to reach.

Yet an understanding of the scientific enterprise, as distinct from data, concepts, and theories, is certainly within the grasp of us all. It is an enterprise conducted by men and women who are stimulated by hopes and purposes that are universal, rewarded by occasional successes, and distressed by setbacks. Science is an enterprise with its own rules and customs, but an understanding of that enterprise is accessible for it is quintessentially human. And an understanding of the enterprise inevitably brings with it insights into the nature of its products.

The Sloan Foundation expresses great appreciation to the advisory committee, now retired. Its members included the chairman, Simon Michael Bessie, copublisher, Cornelia and Michael Bessie Books; Howard Hiatt, professor, School of Medicine, Harvard University; Eric R. Kandel, university professor, Columbia University College of Physicians and Surgeons, and senior investigator, Howard Hughes Medical Institute; Daniel Kevles, professor of history, California Institute of Technology; Robert Merton, university professor emeritus, Columbia

University; Paul Samuelson, institute professor of economics, Massachusetts Institute of Technology; Robert Sinsheimer, chancellor emeritus, University of California, Santa Cruz; Steven Weinberg, professor of physics, University of Texas at Austin; and Stephen White, former vice-president of the Alfred P. Sloan Foundation. Previous members of the committee were Daniel McFadden, professor of economics, and Philip Morrison, professor of physics, both of the Massachusetts Institute of Technology; George Miller, professor emeritus of psychology, Princeton University; Mark Kac (deceased), formerly professor of mathematics, University of Southern California; and Frederick E. Terman (deceased), formerly provost emeritus, Stanford University. The Sloan Foundation has been represented by Arthur L. Singer, Jr., Stephen White, Eric Wanner, and Sandra Panem.

—The Alfred P. Sloan Foundation

Other books in this series are:

Disturbing the Universe by Freeman Dyson
Advice to a Young Scientist by Peter Medawar
The Youngest Science by Lewis Thomas
Haphazard Reality by Hendrik B. Casimir
In Search of Mind by Jerome Bruner
A Slot Machine, a Broken Test Tube by S. E. Luria
Enigmas of Chance by Mark Kac
Rabi: Scientist and Citizen by John Rigden
Alvarez: Adventures of a Physicist by Luis W. Alvarez
Making Weapons, Talking Peace by Herbert F. York
The Statue Within by François Jacob
In Praise of Imperfection by Rita Levi-Montalcini
Memoirs of an Unregulated Economist by George J. Stigler
What Mad Pursuit by Francis Crick
Astronomer by Chance by Bernard Lovell
The Joy of Insight by Victor Weisskopf
Models of My Life by Herbert A. Simon

Preface

How does one interview oneself? How does one achieve distance, reverse one's perspective, see dispassionately that which surges up from
within? If, as my wife Karen says with occasional dismay, I lack ego,
does that make the task easier or harder? Strong egos are easier to limn,
but diffuse egos may be more accessible.

Why does it only now occur to me that, born in 1920, my life is a
defined segment of the twentieth century? That I am a twentieth-cen-
tury man with values, perspectives, even emotions indigenous to, if not
characteristic of, that era just as Samuel Johnson was an eighteenth-
century man and Thomas Huxley a nineteenth-century man? Yet that
thought has only just occurred to me. Perhaps because in the twentieth
century, caught in the swirling tides of change, the future years always
seemed so ill-defined. But now, as it draws to a close, its pattern takes
form, as does that of my life—in either case, doubtless not the form one
might have wished but a form imposed by the logic of chance and
history.

~~~~~~~

Of course, I might well not have lived to my present age. As in every
life, there have been moments of great danger, moments when my survival was highly uncertain and only fate decided. Of some of these, I
was and am sharply aware, of others, perhaps blissfully, not. At least

three such are etched in my mind. Each was mercifully brief, each a confrontation with mortality.

When I was ten, on my way to school, I was struck by an automobile. I was knocked unconscious. Fortunately, the car was light and moving at moderate speed so I was pushed ahead of it some twenty to thirty feet and not run over or crushed. I recovered consciousness in a nearby doctor's office and, in the end, suffered only a slight concussion and severe bruises on the head and leg. What might have been . . . ?

One frigid morning during World War II, we took off from Bedford Airport outside Boston in a modified DC3 with a new radar system on board, bound for Eglin Field in Orlando, Florida. We expected to arrive by late afternoon. The pilot was navigating visually and by radio. I was tracking position easily on the radar.

Suddenly, over northern Maryland, one engine quit. We started to lose altitude. The pilot noted a nearby military airport but, on calling, ascertained that it was closed because of ice on the runway. As we therefore headed toward Baltimore Airport, the second engine quit. Although we had parachutes on board, we were by now too low to use them with safety. The pilot circled back and descended toward the closed airport as our only plausible chance. We landed well on the runway, but braking was to no avail. The plane slid, veered off the runway, and finally abruptly jolted to rest when one wing caught the corner of a building. Although we had strapped ourselves into our seats, we were thrown about by the impact. My head was thrown into the radar gear, which inflicted several cuts and bruises; others were similarly hurt.

We had survived and fortunately the plane could be quickly repaired. The cause of the engine failures was determined to be water that had frozen at Bedford in the lines leading to the secondary gas tanks. A new wing section was bolted on and two days later we were again en route to Florida.

As a graduate student at MIT after World War II, I engaged in several research projects. One concerned measurement of the ultraviolet absorption spectra of nucleic acids and their molecular components at very low temperatures in liquid hydrogen (twenty-one degrees above absolute zero). The liquid hydrogen was contained within a quartz Dewar flask, constructed with plane quartz windows to allow transmission of the ultraviolet radiation. This Dewar was in turn immersed in a larger quartz Dewar, again with plane windows, which was filled with liquid nitrogen to insulate the liquid hydrogen from ambient temperature. The nucleic acid specimens would be immersed in the liquid hydrogen in the path of an ultraviolet beam transmitted through the several quartz windows. While liquid nitrogen was available in our laboratory, liquid hydrogen had to be obtained from a physics laboratory in another building about a quarter of a mile away.

A wooden rack was built to hold the Dewars, which were filled with liquid nitrogen in our laboratory and covered over. I hand-carried the rack to the physics laboratory, where the liquid nitrogen in the inner Dewar was displaced with liquid hydrogen. I then carried the apparatus back to the ultraviolet source in our laboratory. One winter afternoon, I donned a heavy sweater and gloves and set out with the filled Dewars to the physics laboratory. On this day, the path was icy and uneven and, despite care, I slipped. The inner Dewar swung and bumped against the outer Dewar. In an instant both super-cold quartz vessels disintegrated explosively into a mass of minute quartz shards. Incredibly, the force of the explosion was entirely horizontal. Shards of quartz were enmeshed throughout my heavy woolen sweater. I was then twenty-seven. Had the explosion been upward, into my face. . . . Such moments, however banal, speak to each of us of fate, and finitude, and the narrow ledge on which humanity exists.

Conversely, in one's life there are moments of sheer delight when one could wish that time would stop and let the moment endure without end. As I am primarily a visual person, for me these moments come in the presence of surpassing natural beauty and peace, at scenes of wide expanse and steep heights and depth of color.

I remember a May morning along the Big Sur coast, standing amid wildflowers and gazing down on a serene ocean lapping at the rugged shore; a June morning at Rochers-de-Nouy above Montreux, looking down on the shimmering blue jewel of Lac Leman and across to the white slopes of Mont Blanc; the incredible blue of Crater Lake in July, surrounded by pines, still embedded in snow; the shadows and cliffs and rock spires of Canyon de Chelly at golden sunset; the acres of infinitely varied, magnificent tulips at Keukenhof in April, the wondrous soft purple of century-old rhododendron hedges in full bloom in June at Muckross Castle in Ireland. Each of these scenes created feelings of awe and thanks for such a place.

And yes, there have also been moments—days or weeks or months, really—of great anguish. For anguish is a slow emotion, a wound that heals only gradually and leaves an ever-tender scar.

~~~~~~~~

Most of those I knew in my youth are now gone. All of my teachers are retired, if not deceased. Even the structures, the institutions, the patterns of life, if not gone, are vastly changed.

There is a salience to living memory that can never be captured in history books. When I was a boy, Civil War veterans still marched, if somewhat haltingly, in Fourth of July parades. World War I was history to me but freshly alive for the millions who had only a decade before participated in it. And so it is today for the Great Depression, World War II and the postwar years, even Kennedy and Vietnam—events richly alive to me but ancient history to the college students in my classes. In a few more years, all of it will be but dry pages in history texts or grainy films and strangely cadenced tapes.

For years, I have watched uncomprehendingly as my mentors, older colleagues, relatives, friends aged and grew old. Now it is my turn.

For my generation is dying. At the National Academy of Sciences annual meeting, the roll is called of those members who have passed away during the previous year. Formerly, these were the giants of whom I had read but knew only slightly. But now the names are those of my colleagues, those with whom I worked on boards and committees, those with whom I debated or collaborated, exchanged materials, planned meetings, shared evenings and food and drink. Colleagues of the great era of the "breakthrough," when biochemistry and biophysics broke through to the basic level of the gene to find there—in time—the clarifying answers to the enduring questions of biology. Soon the living memory of those events will be gone, as will the living memory of science before World War II, of talent and knowledge freely given to the service of that war, and of the postwar revolution in the conduct of science.

And those of us who yet survive are fast passing into scientific obscurity, our contributions made, we are now irrelevant to the themes of the newer generations of science. "Who? Is he still alive?" Which is as it should be. C. P. Snow wrote, "The scientist has the future in his bones." I once did, in my twenties and thirties and forties. But now my perspectives, my instincts are dated. The future I foresaw has come to pass. A new future awaits and I am much less sure of its form. I read of persons who wish to be frozen in the hope that a more advanced technology will know how to revive them in a century or two. If that should happen, however, they would know no one alive, they would be out of their time, with no shared experiences, wholly adrift, wholly alone.

But I would like to return, briefly, in a hundred years to see how science has resolved some of the questions, the mysteries, with which we wrestle today.

～～～～～

It has long been apparent that we do not enter this world as unformed clay, compliant to any mold; rather, we have in our beginnings some bent of mind, some shade of character. The origin of this structure—of the fiber in this clay—was for centuries mysterious. In earlier times, men sought its trace in the conjunction of the stars or perhaps in the momentary combination of the elements at nativity. Today, instead, we know to look within, seeking not in the stars but in our genes for the herald of our fate. Thus, we are each of us dealt a hand in life through our genes, our early family life, our schooling. It is idle to wish for another. How have I made use of the hand I was dealt?

～～～～～

I am grateful to the Alfred P. Sloan Foundation for having suggested this memoir as part of the Sloan Foundation Science Book Series and for its encouragement and support during the writing.

Introduction

All of life is an emergence out of the inanimate universe. Our very atoms were fashioned in the incandescence of exploding stars. Each human life is likewise an emergence. A new combination of genes is formed and thrust into a particular environment; as it develops, interacts, and learns throughout all the years of life, a new human being emerges. This book is the story of one person, of his continuing emergence, and of his roles in the emergence of a new science and a new university.

I am a biologist, a molecular biologist, I have wanted to understand the nature of *life*, the extraordinary properties of living organisms, and their abilities to grow in a patterned way, to act in a seemingly purposeful manner, to reproduce themselves faithfully and yet to change, to evolve, gradually over myriads of generations. It has been my extraordinary fortune to live in the greatest period thus far in the history of biological science (perhaps it will be seen as the greatest period ever): the era when the machinery of life became evident, when enzymes and genes and viruses became tangible, when subcellular structures, fibers, particles, channels, and vesicles emerged discretely from the obscurity of "protoplasm." During my lifetime, our understanding of the nature of life has advanced incomparably, and it has been my privilege to participate in that achievement.

Scientists find the natural world endlessly absorbing. While they may appreciate the artistry involved, they have no need for the fantasy worlds of the novelist, the playwright, or the movie scriptwriter. The calculus of human emotions, the coils and scars of the human psyche so often mangled in childhood, the thrust and parry, the bonds and fractures of human relations that seem to gratify the egos of so many seem to us often to pale in import compared to the quest for deeper knowledge.

Throughout history, some have sought to live in contact with the eternal. In an earlier era, they sought such through religion and lived as monks and nuns in continual contemplation of a stagnant divinity. Today, they seek such contact through science, through the search for understanding of the laws and structure of the universe and the long quest back through time and evolution for our own origins. Perhaps this urge is a riposte to fate, a nay to human mortality.

But scientists are not monks and nuns. The results of their labors increasingly have major worldly consequences; as Francis Bacon foresaw, knowledge of nature is the key to power over nature. And their "devotions" are costly and require significant public support.

As science has penetrated ever more deeply beneath the surface of matter and of life, the societal impact of its discoveries and their technological applications has become ever more profound. The growth of knowledge has increasingly challenged and taxed our ability to use it wisely. This challenge is all the greater in a democratic society, ultimately reliant on the knowledge and good judgment of all its citizens.

I have throughout my career also been an educator. I have sought to mentor the next generation of scientists. I have also sought through discussion and education to provide that public understanding of science which is essential for its continuing support and its wise use. I have been associated with the two finest institutes of technology in the nation, MIT and Caltech, and was for a time on the faculty of one of the finest—if not *the* finest—agricultural schools in the nation at Iowa State. I was also for a decade the chancellor of the newest campus of the University of California, generally recognized as the finest public university in the United States.

However, even at this level, I have observed little unequivocal success in higher education comparable to that in science. Rather, I have seen education beset with intertwined problems—coping with the great increase of knowledge and the resultant greater demands on the educational process to produce informed citizens and skilled workers; adapting to an augmented democracy that has made higher education a goal of almost all youth with the resultant diversification of the student body with respect to background and motivation; meeting the needs of an increasingly secular society that looks to education to fulfill roles formerly provided by religion. Overall, higher education has lost internal coherence and, with it, external authority. In consequence, it has been increasingly subjected to external oversight and interference.

In my various roles, I have tried to guide this evolving process into meaningful and clarifying directions. But, as compared to that of science, progress in education has been much more tentative, uneven, and difficult to measure. And progress in the scientific education of the general public has been even less able to cope with the pace of scientific advance.

Happily, for some students, higher education continues to provide the doorway to continuing achievement and personal satisfaction. These graduates will be the ones to find the next solutions in science and, one hopes, in education.

~~~~~~~

We live in a unique time in the history of life on Earth, the time when a species first began to understand its origins, its inheritance, its biological functions and their aberrations. After three billion years, in our time we have come to this understanding, and all of the future will be different.

When I entered biological science, it was not obvious that the hidden, mysterious processes of living cells would yield so readily to the tools of modern chemical and physical analysis, that the processes of life would be so linear and so dissociable that individual components could be separated and reassembled to execute individual steps in isolation. In retrospect, we can understand this functional simplicity to be the result of a conservative evolutionary process that made design changes one at a time and that conserved and replicated effective patterns over millions of millennia.

I have been so fortunate to be "in" at this historic unfolding, the "great leap" forward, the penetration of the secrets of inheritance and the machinery of life. But *first* my own life had to unfold. How did it happen that I would be poised to be part of this unparalleled epic of discovery?

The Imprinting

1

Boyhood and Youth—
The Molding

"The past is a foreign land."

A child accepts his world as given. Only much later can he look back and see how the twig was bent.

My mind is a time capsule. A memory is stirred, a scene appears with color, sound, and feel as it was fifty years ago, a half century ago. The others in the scene are dead now, some long since departed. Only my images, my impressions remain—for a time.

I am looking at a small, aging photograph of a boy, about five years old, with a thick mop of hair, dressed in overalls and sandals, walking along a dusty gravel road. He is looking down at the road—pensively? With him is a smaller boy, probably three years old, with short curly hair, also dressed in overalls and sandals, walking along head up, eyes forward.

The older boy is—was—me. I don't recall the taking of the photograph, but I know the setting. Summertime on Washington Island off the Door Peninsula of northern Wisconsin, on a gravel road from the farmhouse to the grocery. Washington Island had no electricity then, no telephones or indoor plumbing. Water came from a pump, ice from the icehouse where it was stored up during the winter and insulated with sawdust. It was one of the few remaining sites of nineteenth-century life, lived according to the rhythms of the sun.

I try to put myself inside the head of that boy, to peel back the layers of years of experiences since accumulated. Of course I can't; the successive years are not simply layered on. They are infiltrated, intertwined, and interwoven into the very nature of one's being. In some deep sense that boy is still here, but I can't "access" him. I vaguely perceive that the world was fresher then, more immediate—the tastes cleaner, the smells more direct, the sights sharper, the sounds more distinct.

Life was still a succession of days of sun or rain or snow, of meals and naps, of games with other children and directives from parents. Compared to the life of today, the life of my childhood seems singularly insulated from outside influence. There was school of course and play-mates, but the home was the primary influence and was little perturbed or violated by the outside world. No TV screens brought distant scenes to our living room. Radio, in its infancy, brought little of interest. Even the telephone, which required a coin for each call, was used sparingly, and long-distance calls were reserved for calamity. The newspapers and magazines—more decorous in that day—brought in the world, but dis-tilled through the flatness of print and the linear, rational process of reading.

We were all less subject to the seductive commercial values of the media and their induced "peer pressures" but all the more captive of the idiosyncratic, sometimes skewed views of our elders. Thus, parents and, later, teachers by precept and example, through word and action and selected reading provided me with a framework in which to order the myriad events of the vast, confusing outer world—a lens selective of importance, a gate sensitive to values. Indeed, the Midwest in which I grew up in the 1920s and 1930s was still very insular. Europe was a week away, Asia two weeks. International commerce was negligible. Issues of global overpopulation, environmental pollution, and inter-national economic management were unimaginable.

When one is young, at least in America, the world is young. The past, all of history, is telescoped. The constraints the past imposes, the hard-won wisdom it contains, the debt we owe our forebears all seem of little moment. Only later, when our lives have been merged into the stream of human existence, do we better recognize the finite scope of our place in human time. I see now that in my earliest years, the 1920s, I was raised in the America of exuberance. The United States was the greatest, the most advanced nation in the world. We had the most ad-vanced technology, the most advanced political system, and newer was always better. Freed of the palsied hand of Europe with its ancient feuds and antiquated governments and frozen social classes, our democracy had liberated the creativity of the people. Our citizens had civilized a continent and created a great industrial society. Secure between two oceans, with no perceived rivals, our destiny was in our hands, and it gleamed. So we thought. Recognition of the side effects of ever more powerful technology or of the social traumas accompanying unlimited free enterprise was yet in the future.

In contrast, my second decade, the 1930s, was gray and grim, a time bleak and foreboding. The Great Depression was psychologically a free fall from the earlier near-euphoria, and the growing menace of Hitler and the war in Europe deepened the gloom.

My forebears on both sides were Germanic and Jewish and thereby melded elements of Teutonic authority and Jewish moral rigidity. My father, Allen, born in 1888 in Chicago, was the elder of two sons. My mother, Rose Davidson, born in 1891 in New York, was the eldest of five children, with two sisters and two brothers. Both of my grandfathers emigrated from Germany to the United States as children with their families, in the 1860s. Their families came for the usual reasons—to escape poverty and prejudice, to seek a better future. Both boys had a limited education; one was for most of his life a salesman in shoe stores, the other a pharmacist who operated a small drugstore. They raised their families in Chicago and New York respectively at only a little above the poverty level, and the education of their children was truncated by economic necessity.

After eight years of grammar school and one year of manual arts high school, my father had to earn a living. After a succession of odd jobs, he discovered a talent for writing. He became a writer for, and ultimately editor of, trade journals. Before and during World War I, he was a feature writer for *Automotive Age,* a magazine for automobile enthusiasts. During that war, the army's use of motorized vehicles was his principal story, so he was sent to Washington, D.C. There I was born in 1920; soon thereafter however, he returned with his family to Chicago, where I grew up. During most of my childhood, he was editor of a trade journal for retail clothing stores.

Despite his limited education, my father was widely read, a self-taught man. Unguided, some of his reading was enlightening and some quite misleading. He was resolved, however, that his children should at least have the opportunity for more advanced education. He sought to encourage intellectual interests by taking us as children on weekends to the Field Museum of Natural History, Shedd Aquarium, and Adler Planetarium. The Field Museum was endlessly fascinating. I particularly remember the huge reconstructed mastodon and the intriguing, yet eerie, ancient Egyptian mummies.

After three years of high school, including a final year of secretarial training, my mother was similarly obliged to go to work. However, she intensely disliked office work. She was a very pretty woman and met my father on one of his trips to New York. They were married in 1913 (he was twenty-four, she twenty-one). My older brother, Allen, Jr., was born a year later. In Germanic fashion, my father dominated the household. With three sons (my younger brother, Richard, was born two years after me), my mother played a relatively passive role in the household, occupying herself with projects at the synagogue and social activities such as bridge and mah-jongg.

In our home, there was a strong emphasis on intellectual and cultural development, but at a cost. Emotions were not to be trusted. Emotional display, except through such refined media as music or art, was to be repressed as crude, primitive, and unworthy. A distant mother and the absence of sisters reinforced this atmosphere. We learned to be self-reliant, to make our own way in this world. We could expect some support from the family but not much else. The anguish of the Great Depression confirmed this view as did, later, the "sink-or-swim" attitude prevalent during my years at MIT.

I was also brought up with a strong sense of duty—with the charge to use my talents, which it appeared early were considerable, for the benefit of others in whatever way seemed at the time most propitious, and not to be distracted or seduced from that obligation by transient pleasures. At the same time, my upbringing oddly provided rather little sense of community or belonging to an ongoing stream of human life and endeavor; this has only come to me much later in the global community of science. I was brought up not to expect much from life unless I "earned" it. Perhaps this attitude was simply in the American tradition of "rugged individualism." Perhaps it was the lost contact with our ancestors, buried far away in a decadent Europe from which we were thankfully delivered. Perhaps it was our Jewish identity, which we understood excluded us from the American mainstream. No Jewish boy would ever be president, head of a great corporation, or even mayor of Chicago. America was much freer than Europe and many professions were open to us, but the exclusionary pattern was still there, still real and strong. A Jewish boy would have to be better, we heard, as we hear today for a woman, an African American, or an Asian American.

Much later, I came to realize that, in part, these perspectives reflected, beyond any intrinsic merit, my father's psychology, even pathology. He was a deeply fearful man. He was beset by, and transmitted to his children, the convictions that life was a hazardous enterprise, that one could easily make an irreversible mistake that would forever blight one's health or economic or social status, and that the path to avoid such disaster was one of constant caution, moderation, and modesty in behavior and action. Not to dare greatly, not to plunge wildly. He was highly conscious of appearances and highly threatened by possible loss of dignity or community censure. To become drunk, for instance, was not merely an immoderate act but a disgrace, a blot on one's image.

To me, growing up in his household, all of this seemed reasonable and proper; questioning of parental authority and wisdom was not encouraged. And the apparent correctness of the basic proposition was certainly corroborated by the external events of the 1930s. The Great Depression and the looming abyss of World War II reinforced the concept that one's individual destiny was subject to great and perilous forces far beyond one's control. So it was not until many years later that I came to realize how skewed, pinched, and constrained a view of the world this was and to escape, at least in part, from its thrall.

I am left-handed. Fortunately, my father had read of the traumatic consequences of forced conversion from left- to right-handed use and intervened at school to prevent such a requirement. However, for many years I thought I was clumsy because I had much more difficulty than others with scissors or screwdrivers (and later, corkscrews).

Schoolwork was ridiculously easy for me. I passed the first two grades in one year, likewise the fifth and sixth. Of course, this advancement into an older age group distorted and retarded my social development. Because of this, and because school was mostly boring, I took refuge in reading. I certainly read the majority of books in the children's section of the local branch library.

In my teens, two books markedly changed my view of biological life. In my family, life in the biological sense (much less sex) was rarely discussed. It was a given fact of nature not subject to analysis or human understanding. Our bodies were bequeathed to us and we had limited control over their subsequent fates. But from Wells and Huxley's *The Science of Life* I first realized that one could regard living organisms as very complex machines, with varied components whose functions and interactions could be dissected and analyzed. Of course, we have now carried that perception down to within the cell, even to the genes, and have found machines within machines within machines to the limiting level of molecular devices.

Except in the crudest sense that like begets like, my upbringing provided no appreciation of heredity. Thus, the book *You and Heredity* by Sheinfeld was a stunning entry to a new world. We are what we are by virtue of our specific, inherited genes. Of course! But how? What wondrous processes could produce this result? Not immediately, but subtly, these new insights gathered deep within me and generated powerful fountains of interest and curiosity that have continually renewed and refreshed a lifetime. Indeed, I seem to have been endowed with an unending curiosity and I have always found that my interest in almost any subject grew in proportion to my knowledge about it. But biology has had a first claim.

Although the youngest, I was the valedictorian of my grammar school class and went on to high school, a larger world of some five thousand students. At this same age, several other influences affected my life, among them the Great Depression, boys' camp, Sunday school, the World's Fair.

Although the Depression heightened my father's fearfulness, it did not in fact greatly alter our economic status. My father had always lived well within his income and, while his salary was reduced, he was never unemployed and our standard of living changed little. But I well recall grown men coming daily to the back door of our apartment to beg for food; we never turned them away empty-handed.

And I recall a favorite uncle who arrived at our door at a very early hour one morning because he literally had no money to buy food for his family. My father helped out for a time. These images of despair remain sixty years later.

For four summers, my brothers and I went to a boys' camp in northern Wisconsin. Here I was introduced to a more elemental, more natural world. Northern Wisconsin is dotted with clear lakes set in pine forests. In those days, before motorboats and water skiers, the lakes were extraordinarily quiet and tranquil. Paddling across in a canoe, I could imagine a kinship with the Hiawatha of legend.

Camp was also my first interaction, in many aspects of living, with a considerable diversity of other children, many from families much less repressed and with values quite different from my own. The experience was illuminating, although insufficient to raise strong doubts about parental strictures.

I was also introduced to athletics, especially baseball and tennis for which it turned out I had some aptitude and which I greatly enjoyed. Baseball was softball and I pitched reasonably well. I became a fan, following all the major league teams and players and devouring statistics. Baseball has remained a lifelong passion, although opportunities to play were always limited by the need to field eighteen players. Tennis became a more readily supported addiction, and I spent countless hours after school and in the summer on the local public courts. If no opponents were about and a court was available, I would practice serving by the hour. I became a good player.

My parents belonged to a reform Jewish synagogue and I had attended Sunday school from an early age. But now, as confirmation age approached, the lessons and sermons took on more significance. The tenets of reform Judaism are morally lofty but leave little scope for human frailty. With its insistent command to consider all the consequences of one's deeds, it was not a creed to encourage spontaneous action or uninhibited emotion.

A World's Fair was held in Chicago in 1933 to celebrate the city's hundredth birthday and was extended to 1934. Its theme was "A Century of Progress," and it featured many exhibits of the latest science and technology. I and a school chum, Jim Flood, went frequently each summer. The advances displayed in science and transportation and communications were exciting. I remember particularly an exhibit featuring samples of all of the chemical elements that had then been isolated, arranged in a periodic table; also a fascinating display of human embryos at the various stages of development. But, surprisingly, the most memorable event was my discovery of Shakespeare. An Elizabethan theater featured a replica of the Old Globe, with a repertory company that daily performed Shakespearean plays. I was mesmerized not so much by the plots, which seemed archaic, even contrived, but by the language. I had never heard such beautiful and poetic language used to express profound thought and emotion. This response was reinforced later in high school when we read Shakespeare. To this day, I am enraptured and inspired by Shakespeare's speech, so soaring and lilting and at the same time so replete with meaning.

Our genes provide our physical frame, much of the specific basis for our personality, and the raw material for our intellect. Circumstance, environment, and culture, however, map the specific routes for intellect.

In high school, I had two extraordinary teachers, Miss Shoesmith in mathematics and Mr. McClain in chemistry, who surely influenced my future. Miss Shoesmith inspired me throughout geometry, trigonometry, and college algebra through the use of "special credit" problems, all beyond the regular class assignment and of increasing difficulty.

These interested and stretched my mind and, in a subtle way, generated a growing capacity for innovative problem-solving. Moreover, I enjoyed the challenges.

Similarly, Mr. McClain, by letting us perform "extra credit" experiments in the chemistry laboratory, expanded and strengthened my facility with laboratory apparatus and my capacity to plan an experiment and record and analyze the outcome. One Parents' Night, he let us design a set of demonstrations using liquid air, involving such oddities as a hammer made of frozen mercury and a small steam locomotive that ran by the expansion of vaporizing liquid air.

As a teenager, I discovered the joys of science fiction and eagerly awaited the next issues of *Astounding Stories* and *Wonder* magazines. The science fiction of that era was thin on plot and characterization and emphasized technological extrapolation, much of which has, in fact, come to pass. Science fiction today is more literary and indeed much of it is social-science fiction, based on extrapolation of one or another societal facet.

In senior English, in the spring of 1936, the principal assignment for the first semester was the preparation of a lengthy theme, involving library research, on some significant topic. With seeming prescience, I chose "The Transmutation of the Elements." My theme covered the older, misguided efforts to achieve transmutation, doomed to failure by ignorance of the basic nature of the changes required. After this, I discussed the current understanding of the nature of the atom and the atomic nucleus, the achievement of transmutation on a minute scale by nuclear physicists, and the potential source of energy locked in the nucleus. I anticipated that, some day, large-scale transmutation and large-scale energy release would indeed be possible. I did not imagine that day would come within but a few years, to change all our lives.

In grammar school, at about seventh grade, I had become aware that I was, in public, painfully shy. To be called on to speak before the class— or worse, to read to the class an essay I had composed—was an agonizing experience. Yet, from my parents, and somehow from within, I knew this experience could not be averted. In future life, it would be essential to be able to make known my views by speaking in public. So in high school I enrolled in a yearlong course in public speaking. This was learning by doing, for me by ordeal. We were taught some rudimentary techniques of public speaking and even a few acting skills, but the important lesson was simply the conditioning, the confidence gained by repeated public presence without resultant censure. Yet even today, while a technical seminar or lecture on a scientific subject is no problem for me, a speech expressing personal views on a controversial subject generates a tautness, an anxious tension.

I have always been a future-oriented person, a fact that causes some internal conflict now as my future grows increasingly finite. In part, this orientation may have arisen from, or perhaps accounted for, my interest in chess. Chess, of course, requires that one plan and anticipate the consequences of one's moves for several steps ahead. I first started chess with my father, at about age ten. Then, he could defeat me easily and would give me a piece advantage to make the game more even. After a time we became more equal. In high school, I joined the chess club and, with this exposure to other players, soon surpassed my father, a small but psychologically important step. I also began to read chess books. Subsequently, I made the school chess team and in my last year, I was first board. Our team won the Chicago high school city championship. Later, in college, I simply lacked the time to continue with chess; since, I have played only sporadically.

This future orientation likely also derives in part from my rearing in the Jewish tradition with the sense that one should not merely while away one's time here on earth; that one should contribute to and be part of a more enduring human enterprise. And so my life has been spent in education and research, the two ways in which our society most directly invests in the future.

When I graduated from high school, I was not quite 17. Because I was a year younger than most of my classmates and was regarded—to varied reaction—as a "brain," my social development, and particularly my relationship with girls, was quite retarded. It was my serious misfortune to have no sisters. With only brothers and a remote mother, the world of women—their desires, needs, interests, and goals—was foreign and obscure to me and, despite considerable interaction since, has in good part remained so.

It was time to think about college. My older brother had gone to the nearby University of Chicago to study law. My interests more clearly lay with mathematics and science. My mother, reared in New York, had long believed that the Eastern schools were superior to those of the Midwest, and she felt that my talent merited the best education. Stan Jarrow had been my laboratory partner in chemistry and we were good friends. He had his heart set on an engineering education at MIT. MIT, or Boston Tech, as it was sometimes called, was held in high repute even in Chicago. It seemed an apt match to my interests, which at the time leaned toward chemistry.

But we knew no scientists. I had no role models.

My father had a limited understanding of science and was unsure about the employment prospects for chemists. Their contribution to, and therefore worth in, society was unclear to him. Even a conversation with Mr. McClain did not help. Chemical engineering, however, seemed to him to be a more practical subject, so it was agreed that I should apply to MIT to enroll in that field.

This was 1936, in the depth of the Depression. Living away at MIT would manifestly be more costly than living at home where I could attend the University of Chicago. But my father agreed that if I could receive a scholarship from MIT (in those days scholarships were primarily merit-based), he would send me there for at least one year. While I was not the valedictorian in my high school class, I was the highest ranking male and am sure I received good letters of recommendation from my teachers. I was awarded an MIT scholarship for full tuition at $500 per year.

This lad with my name, age sixteen, steadily gazing out from his allotted square in the high school yearbook is closer to me today but still indistinct. His features are recognizable, but he himself is yet only half-formed, half-educated, half-emerged from the parental cocoon, still trailing wisps of old myths and superstitions and prejudices, with views not yet his own, skills partly honed, perspectives short and fractured, but an absorbent mind and an endless curiosity. Sure, but quite unsure, and somehow imbued with the stiff determination and internal discipline to "make it," to succeed, wherever the amorphous future would lead.

In 1932, M. Knoll and E. Ruska invent the electron microscope, which extends human vision to the submicroscopic and, in time, to the macromolecular level. Viruses can be seen and essential structures and processes of living cells revealed.

Also in 1932, Curie and Joliot discover artificial radioactivity, the production of radioactive isotopes not found in nature. The use of these isotopes as tracers, substituting for the natural atoms, has been essential to the elucidation of biochemical pathways and structures.

In 1935, Wendell Stanley produces crystals of the tobacco mosaic virus, leading to the possibility of subjecting viruses to detailed physical and chemical analysis.

2

Transition 1

Leaving home, truly leaving home, for the first time was a wrenching experience. I had done so before when I attended summer camps, but this was unmistakably different.

It was time. I was becoming increasingly restive under the rigidities imposed by my father's fears—I did not yet realize how much I had by then internalized. But to leave family, friends, the familiar neighborhood and city, the grid of school and stores and movies and sport facilities and transportation systems for a blank—Boston, a dot on a map—was daunting.

I did not then realize that, in fact, I was leaving home forever. Except for a few weeks during college years, in summer or at Christmas, I would never again live in Chicago. My high school friends would tread different paths. Even my place within the family was henceforth to be different, separate, increasingly external, even as the family I had always known lost its structure.

Where was MIT? We had been told it was on Massachusetts Avenue, just across the Charles River. Now we were on the bridge. It couldn't be that six-story red brick building on the left, almost surely a hotel. Was it that massive grey structure on the right, looking more like an industrial plant? It hardly resembled a university campus. As we drew closer and the Great Dome came into view, I recognized the scene from the MIT catalog and realized the truth. Here, for better or worse, was my future.

My parents had driven with me to Boston to start me off at MIT— a three-day journey in those days. We now located Bemis Hall, the oldest dormitory, where I was to be in room 413. The room was small and sparsely furnished. A bed, a bookcase, a desk and chair, a reading chair. No lamps. A sink and mirror, a small closet. Showers and toilets down the hall. One window looking across a court to another dormitory.

The next day my parents went on to New York, and I moved in. I acquired a floor lamp from the porter, who had a small side business of collecting floor lamps at year end and reselling them the next fall. That evening I looked about the spartan dormitory room with an intense churning mixture of sadness, uncertainty, and eagerness. Sadness because I knew in some visceral way that I was now outside the nest. Home—the only one I had ever known—was a thousand miles away. Uncertainty because an unknown world was all about me and before me. How would I stand up in this strange setting? Of course, I was not wholly on my own, but I had never been quite so much on my own. It was now up to me. And yet eagerness—to explore, to enter the world of science and technology at whose door I literally now stood. In those formidable buildings across the way was the gateway to my life.

3

MIT—The Shaping

"MIT is a place for men to work and
not for boys to play."

MIT shaped my mind.

Strangely, it began with a rather farcical freshman camp, off in the woods of Massachusetts. The camp was in some ways a hoax: instant, and brief, camaraderie with faculty and alumni. At a faculty-alumni baseball game, President Karl Compton pitched—I never saw him again until graduation. We heard lectures about the history and traditions of the institute, about its expectations of its students: "MIT is a place for men to work, not for boys to play."

Indeed, MIT was a stern place. Student counseling, psychological help, tutorial sessions were nonexistent. The academic pace was swift and unrelenting. It was very much "sink or swim" save for what mutual care and assistance the students could provide each other. I was very interested in my 625 fellow freshmen, almost all male. While women were, and always had been, admitted, they then made up 1 percent or less of the student body. Most of my classmates were from the East— New England, New York, New Jersey. Many were from New England preparatory schools or elite public schools such as Boston Latin or the New York Academy of Science. A significant fraction were from abroad. Would my Chicago preparation be adequate?

MIT provided a very special kind of education that left its imprint on all who attended (and succeeded). It gave us the experience of working "flat out"—at our full capacity, for extended periods—which produced a realization of our capabilities and a confidence in our competence. It taught us, by daily repetition, the art of problem-solving: how to frame a problem, how to find a feasible approach, how to bring to bear any or all of our skills and knowledge upon its solution. This became a life-long habit. It ingrained in us the importance of quantitative thought, the value of a feeling for orders of magnitude, for exponential processes, for precision within "significant figures." This fit us well for the quantitative, practical worlds of science, engineering, and economics, perhaps less well for the qualitative worlds of art, literature, values, and political maneuver.

MIT was little afflicted with self-doubt. It had a clear and distinct sense of its pedagogic mission, rare in academia: technology was the future and it was good and MIT was its leading incubator, a beacon for civilization. And the institute transmitted this sense of worth to its students.

In my day, all freshmen took the same curriculum and much of the second year was common. Two years of physics, two years of calculus and differential equations, one year of chemistry, two years of English and humanities, two years of ROTC, one year of mechanical drawing, one year of physical education or athletics. Laboratories or drafting every afternoon to be followed by an hour of ROTC. Problem sets every night. An hour quiz every Friday morning, alternating between mathematics and physics. An unrelenting pace.

Overall, my preparation proved inferior to that of the prep school and Boston Latin graduates. My background in chemistry and mathematics was adequate, but English was deficient and physics was abysmal.

I have always been grateful that MIT required this broad and deep education in the basic sciences and mathematics. For while I have made limited direct use of many aspects, this background has enabled me to follow with interest and understanding the remarkable developments in physics, chemistry, and astronomy over the decades—an outcome that has expanded my range of colleagues and enriched my entire intellectual life.

Faculty were both awesome and remote to freshmen, yet they had recognizable human foibles. The freshman class was divided into twenty-five sections of twenty-five students each. There were discussion (problem-solving) sessions in physics and chemistry for each section in addition to the large lectures. Professor Van de Graaf was assigned to teach our physics section. At that time, he was hard at work on his famed electron accelerator. Often on Monday mornings, he would come to class a bit bleary, obviously having worked all weekend on his research and not having looked at the assigned problems. While they were problems of beginning physics, sometimes they were not all that simple, even for him. Chemistry and mathematics came rather easily and obviously to me, but I soon found I had rather little intuitive feel for physics, at least for mechanics and heat, and I had to learn these as abstract, quasimathematical science.

English composition almost drove me from MIT. Someone—I was told it was Vannevar Bush—had introduced the notion that a good way to teach composition to MIT students would be to require them to write descriptions of common objects, as for a patent application! This is a formidable task. The objects assigned included a Stilson wrench and a plain, brown-paper grocery bag. These assignments were fiendishly difficult and, to me, basically uninteresting, but they did engender a lasting respect for patent attorneys.

Freshman life at MIT brought varied experiences and required frequent adaptation, even though schoolwork consumed most of the available hours. Freshmen were hazed in those days. They had to wear a special tie in institute colors (actually a good idea as it permitted instant recognition of other freshmen), to run errands for upperclassmen, and generally to be available for sessions of harassment that frequently ended in an enforced trip to the showers. All of this came to a climax on Field Day, immediately prior to Thanksgiving. This event involved several athletic contests between freshmen and sophomores, including relay races and a tug of war and ending with a "glove fight," a more-or-less genteel form of mayhem. All freshmen and all sophomores participated, each class wearing a distinctive glove. Whichever side could get the most gloves off members of the other class and behind their own goal in the time allotted won the match.

If the freshmen won Field Day, hazing ended. Otherwise it continued to the end of the first semester. Being more poorly organized, freshmen rarely won, but, remarkably, our class ('41) did.

Freshman year provided many new experiences, pleasant and unpleasant. Lacking female students, social life at MIT was difficult. "Mixers" were arranged for freshmen with freshwomen at local women's schools such as Simmons College. These were awkward affairs that seldom led to continuing relationships, although it was refreshing to talk with young women who had no preconceived impression of me. Fortunately, in time we came to know fellow freshmen from the Boston area who had networks of female acquaintances. This led to dates and friendships.

The Thanksgiving holiday was too brief to return to Chicago, so we remained at MIT. Eating Thanksgiving dinner in a restaurant seemed to me joyless—and still does.

Several students in the dormitories, of German descent, were open admirers of Hitler, had swastika flags and emblems, and affected storm trooper garb. This surprised me and caused several near confrontations.

A group of Cuban students reacted to their first snowfall with palpable delight, rolling in the snow, hurling snowballs, and making snowmen.

Boston, prior to the post-World War II renovation, was yet the old city. From across the Charles it still had a European skyline—no skyscrapers, buildings limited largely to walk-up heights. Historic squares such as Haymarket, Scollay, and Copley retained their old character. Compared to those of Chicago, the streets (old cowpaths, it was said) wound, curved, and changed names in midblock. Much of the city— Chelsea, Somerville, Roxbury—seemed old and beat-up. The area of Cambridge behind MIT was literally a slum. On the other hand, Beacon Hill was charming, as were suburbs such as Brookline, Newton, Lexington, and Milton.

At Christmas I went "home," but it was now no longer home. My bed and bureau were still there, but now I was a stranger: not quite up on family events of the recent past, not included in future family plans. But more significantly, I had entered into a world of science that was foreign to my family and that, sadly, I could not bring home to them. The language was too strange, the concepts too abstract, the images too unfamiliar. While they approved of my direction, the journey could only lead to further separation. I saw old high school friends, and here too the divergence was evident—we were on different paths, and I would not return again for seven months. A long time at eighteen.

We anxiously awaited our first-semester grades. While I felt I had done well, the finals had been difficult, and they counted heavily. In the end, all went very well and I was truly relieved. I could do it. I could compete with the best at an elite, tough school.

At the end of my first year, I stayed on for six weeks to take qualitative inorganic analysis. The technique then was based on the use of inorganic sulfides. So we spent all day, five days a week, in a laboratory reeking of hydrogen sulfide, to which one soon became so habituated that it had no odor (except to your friends when you came out of the laboratory).

When I returned home again to Chicago, I was exhausted—by the long year and, I suspect, from the sulfide exposure. I did little but rest for a few weeks. My parents and family (both brothers still lived at home) had moved that spring to a row house near the University of Chicago. It was pleasant but more than ever not my home. My former close friendships were now very tenuous. Even those friends who had remained in Chicago had gone their separate ways to different jobs or different schools.

As I traveled back to MIT that fall, the great hurricane of 1938 swept through New England, uprooting one third of the trees, devastating the shoreline, flooding rivers, and destroying bridges. My train, bound that night for Boston, was rerouted to New York. In New York, no trains were leaving for Boston; the only access was by air, so I had my first airplane ride on TWA from Newark to Boston. I can recall the sensation of being airborne: the sudden absence of jolting, the roar of the engines, the lights of Boston coming into view, the thump of the landing. Boston was hard hit—large sections had no phones or power. But MIT was mostly unscathed.

Sophomore year began my chemical engineering curriculum, and by second semester I was having serious doubts about this choice. I was simply not that interested in pragmatic solutions to practical but often intrinsically trivial problems. I preferred to search for deeper knowledge. I was seriously considering physics when I read in the school paper of plans to revise and revitalize biology at MIT. The seeds planted by Wells and Huxley's *Science of Life* suddenly germinated.

Biology at MIT had long been a service program, originally for a course in sanitary engineering, arising out of civil engineering. Later, a program in public health had started at MIT in the 1890s. This program had been important in the first quarter of the century, but by the 1930s the scope of the field had expanded, and an M.D. was now required for all important positions.

MIT decided then to phase out this program and to replace it with a program in those areas of biology, biochemistry, and biophysics that could be complemented by other strengths of the institute. They brought Francis Schmitt, a well-known biophysicist from Washington University interested in cellular ultrastructure and nerve function, to be chairman and to recruit a faculty. They also brought John Loofbourow, a physicist from the University of Cincinnati who was interested in the effects of ultraviolet irradiation on cells and who had recently written a comprehensive review of biophysics. They were establishing a five-year program, leading to a combined S.B. and S.M. degree in quantitative biology and also (how presumptuous, prescient, and premature) in "biological engineering," a name that perished after a few years—it was four decades too soon.

I decided to transfer to this new program. What made it so intriguing? As a boy I had read avidly of the great explorers—DeSoto and LaSalle, Coronado, Lewis and Clark. Men who had explored a whole new continent, who saw new worlds and made major and permanent additions to both the global map and the sum of human knowledge. Discovery in geography was now largely complete, but not in science. True, physics and chemistry seemed well established—by no means completed (indeed, synthetic chemistry seems inexhaustible) but in outline well mapped. But biology, beyond the descriptive, seemed terra incognita. Here was the unknown continent within the living cell, and here were new tools, new approaches to the deep and mysterious processes underlying life—and therewith, ourselves.

I was somewhat apprehensive when I broached this subject with my father, remembering his initial concern about a choice of chemistry, but he fully supported me. As I had shown that I could do so well at MIT, he believed that I should be given the opportunity to do what I really wanted to do. Science was becoming a way of life for me. While he could not comprehend or follow where I was going, he admired the direction and respected my choice.

That summer I again stayed on for six weeks to take a course in qualitative organic chemistry. Once again, it was five days per week in the laboratory. The course was enjoyable. Mostly, it consisted of devising procedures to identify unknown organic compounds. This was an excellent means to require the students to learn and think through the application of analytical procedures as well as to become familiar with the great handbooks of chemical compounds, such as Beilstein's *Handbuch der Organische Chemie*. Some of the compounds were a problem to me because I was sensitized to them to a degree that they caused rashes on my hands. In retrospect, I feel sure that at least several of the unknowns, such as nitroso compounds, were mutagenic and carcinogenic, but in those days we were innocent of such concerns.

All MIT students were required to take a year of economics. I had taken the first semester in the spring. The theory had seemed mathematically rather simple and of limited applicability to the real world of business. I decided to take the second semester by examination, just prior to the fall semester. Regrettably, during the summer I did not find time to read the textbook. As fall approached, the only solution seemed to be to return to MIT a few days early and study. I did so, and in two days I had read and absorbed the text. I took the examination and passed with a B grade. This is not a recommended way to learn a subject. But as a means to fulfill a requirement, I have seen it performed by many, many students in the years since.

My transfer to the new biology program was premature for it was not yet in place. Many of the new faculty would be arriving during that year or next, and the new courses were not yet available; only the older biology "service" courses, many antiquated, were taught. Thus, the course in bacteriology was concerned not with microbial physiology but with the identification and classification of organisms (as appropriate for public health students). The course in invertebrate biology was largely an exercise in the microscopic dissection of various worms and insects.

I was also taking physical chemistry and atomic physics. The physical chemistry was, unhappily, a classical course more suited to chemical engineers than to biologists. Atomic physics was my introduction to the world of quantum phenomena.

Absent new biology courses, I took a variety of potentially valuable courses including crystallography, advanced optics (all geometric in those days), advanced organic chemistry laboratory, and advanced microscopic techniques. A particularly interesting course in X-ray diffraction was taught by Professor Warren. Though the course was largely limited to inorganic crystals of varying complexity, there was some discussion of early work on organic molecules. Dr. Fankuchen, who had recently returned from a year in Bernal's laboratory in London, was spending the year at MIT. I had some discussion with him concerning the possibility of applying X-ray structure analysis to proteins, some of which had then been crystallized. While the potential was clearly present, the difficulties of data collection and analysis seemed overwhelming.

I was greatly intrigued at that time by Vannevar Bush's development (at MIT) of analog computers to carry out differentiation, integration, and numerical analysis. Interestingly, only the future development of digital computers made feasible the X-ray diffraction analysis of protein structure.

A course in statistical analysis, especially as applied to small sample bases, proved valuable in providing me with a good understanding of statistical variation and its importance in the analysis of data. It also gave me an appreciation of the difficulty and tedium of performing computations and analysis with large data bases using the mechanical calculators then available.

I was taking many interesting courses out of the wealth that MIT offered—but I was not learning much biology.

In biophysics, Professor Horton from the electrical engineering department had become interested in the electrical properties of various tissues at different frequencies and had joined the new biology program. Under his influence, I took the electrical engineers' year course in electric circuits. This was a rigorous, quite mathematical course in circuit theory that provided a thorough grounding in the use of vectors in the imaginary plane to analyze alternating current problems. I also took a laboratory course in electrical measurement—this training in electrical science was to prove useful in a very unexpected way. Professor Horton also ran an electronics laboratory for biophysics students in which we constructed an electronic pH meter, an advanced instrument in those days.

We had a year-long course in biochemistry, along with courses in enzymology and animal physiology. It is difficult today to realize the primitive state of biochemistry at that time despite the fact that the subject dated back to the middle of the nineteenth century.

The living cell was still an object of mystery, a "black box" that performed remarkable feats by unknown mechanisms. Under the microscope, cells could be seen to move, to grow, to divide and thus multiply. Complex movements and rearrangements, called mitosis, accompanied cell division. But clearly much of the mechanism lay in structures and reactions invisible in the light microscope.

We knew that cells required a source of energy: light for those that could perform photosynthesis, nutrient for others. Cells used molecules of nutrients for growth, during which these were digested and converted into more cellular substance. Specialized cells in the body could contract and perform work, or conduct electrical signals, or detect light, or secrete a wide variety of substances.

The major classes of the chemical constituents of cells were also known: the proteins that catalyzed the chemical reactions and that formed major structural elements, the carbohydrates and the lipids, the obscure nucleic acids, and a wide variety of smaller molecules. The pathways of degradation of large nutrient molecules into fragments (inter-mediary metabolism) were partly known. The processes by which these fragments were synthesized into the larger molecules characteristic of each cell were completely unknown.

The chemistry of inheritance was a profound mystery. Genetic factors could be transmitted unchanged for many generations, or they could mutate and the mutant form could then be similarly transmitted un-changed for many generations. Sometimes the mutant form could re-vert to the normal version, or it could mutate further. The nature of the genetic factors and their mode of action were completely unknown. What kind of chemistry could account for these observations?

While the stages of embryonic development of various organisms had been described morphologically, the reactions and processes underlying the development of an adult organism from a fertilized egg were com-pletely obscure. Clearly, intricate patterns of control existed: internal control of synthesis and transport within each cell and "social" control, coordinating the actions of cells within an organism. But their nature was totally unknown.

There was much to learn before biology became a deep science, based on broad general principles. To believe that these complex, spon-taneous, near-miraculous processes could—and would—be explained in terms of chemistry and physics required an act of faith. To undertake to find such explanations was an irresistible challenge.

By 1940, it was well recognized that most reactions in cells were specifically catalyzed by enzymes. It had finally been settled that en-zymes were large molecules, proteins, and not some mysterious "vital force." A few enzymes had relatively recently been purified to a degree from which they could, with difficulty, be crystallized. Crystallization of an enzyme was an important part of the laboratory course. However, while it was known that proteins were made largely of amino acids, no one knew the amino acid composition of *any* protein, much less the sequence or the spatial disposition of its amino acids. The mechanisms of enzymic catalysis by proteins were therefore completely obscure. Even the molecular weights of the best characterized proteins were known only with considerable uncertainty.

Indeed, this uncertainty permitted the formulation of daring but quite erroneous hypotheses, such as that of Svedberg that the weights of all proteins were multiples of a basic unit. Or that of Dorothy Wrinch, who sought to account for the presumed regularity of protein molecular weights on the basis of a regular crystallographic structure for all individual protein molecules that could be scaled to accommodate the various weights. An ingenious theory, but totally wrong. Nature has its own conception of ingenuity.

No one had any clue as to how the proteins—or most of the components of cells—were synthesized. One problem was scale. Only a few proteins or other molecules could be obtained in quantities suitable for the techniques of organic chemistry. Tracer techniques with radioisotopes and highly sensitive assays were in the future. Most enzymatic reactions were followed either colorimetrically (with human eye detection) or by coupling to some reaction that permitted a gas to be evolved or absorbed. The volume of gas thus affected could be measured, very tediously, in Warburg manometers.

We spent a moderate amount of time in the course on the origins and interconversions of various mold pigments. I ventured the opinion that this did not seem like a central problem in biochemistry. Unfortunately, I thereby unwittingly seriously affronted my instructor, whose research was in that field.

Research in biochemistry then was very individual and small scale. No general external sources of funding were available and, in the continuing economic depression, institute resources were very limited. I recall my enzymology instructor telling us how he had had to defend a request to the dean for some one-milliliter pipettes. The dean knew that he already had some graduated ten-milliliter pipettes and did not see why the one-milliliter size was also necessary.

Some of the exciting developments in biochemistry at that time concerned vitamins, viruses, and the beginnings of research on antibiotics. Vitamins (essential nutrients) had been discovered in earlier decades, but additional factors were being discovered and chemically identified. Kögl, in an heroic effort, had isolated a sufficient quantity of biotin (vitamin H) to identify its chemical structure.

The nature of viruses was more obscure. Wendell Stanley had recently startled the field by producing quasicrystals of the tobacco mosaic virus, leading to the hope that these mysterious entities might now be characterized in physical and chemical terms. At the same time, reports of antibacterial compounds from molds were beginning to appear in the literature; penicillin and gramicidin were among the first of the antibiotics.

This was a more leisurely time in the adolescence of the science—*Chemical Abstracts* was published biweekly, and one could go to the library and literally read through the entire biochemistry section in a couple of hours on a Saturday morning. The *Journal of Biological Chem-*

istry, the leader in its field, did not publish in July or August—the editors, and presumably the readers, deserved a vacation.

In 1940, biological functions were simply observed phenomena that could not be related to underlying structures because of the basic lack of structural information regarding macromolecules or subcellular organelles. The sharp contrast between the rigorous and logically exclusive mathematical analysis of my physics, statistics, and electrical engineering courses and the qualitative and often tenuous experimentation in biology—relying on intuitive art as much as logical projection—disturbed me. Very different intellectual approaches were required for the different subjects. My aim would be to advance biological science toward the levels of understanding achieved much earlier in the physical sciences.

All of this highly technical education, both physical and biological, may seem to have provided a one-sided, if rich, intellectual development. It did.

Not that I had no interest in literature, history, philosophy, or biography. The opening world of science at MIT was so vibrant and alluring, the program so intense, the faculty, especially in science, so competent and eminent, and the value structure so oriented that the more "humane" fields were simply crowded out, ostensibly "deferred." There was a small humanities-type library in the Walker Memorial in which I would very occasionally browse. In particular, I remember being fascinated by a world history written from a perspective different from any I had encountered thus far—that of Jawaharlal Nehru while in prison.

Today, five decades later, MIT is significantly different. Three times as many undergraduates and an equal number of graduates throng a much expanded campus. Thirty-four percent of the undergraduates are female, and fourteen percent are underrepresented minorities. Tutorial assistance, psychological counseling, and other services are readily available. The course offerings in the humanities and especially in the arts have been greatly expanded and diversified. Topics related to the interactions of science and technology with society are prominently discussed. An extraordinary abundance of student organizations offers opportunities to engage in all manner of non- or quasi-academic activities.

As might be expected, this wide diversification has come about at a cost. The depth and intensity of the common basic science education has diminished to accommodate the greater breadth of the curriculum and the greater range of interests of the student body. Today's MIT students receive a broader education that may well better prepare them for the complexities of today's society but that narrows and dilutes their scientific base.

All was not equations and test tubes. In my senior and fifth years I roomed with Art Graham in the Senior House. We shared a suite—a living-study room, a bedroom, and a dressing room. The arrangement of four suites to a floor provided environs a bit more gracious and quieter than had the long dormitory corridors of the three previous years. We looked out over the President's House and the Charles River.

Taking an overload as usual, I was still intensely engaged in studies, but in general it seemed to me that we seniors did not have to work quite as hard as in earlier years. Also, by then we were well acquainted and there was considerable camaraderie. There were long evening discussions about science, about the approach of graduation and the life after that, and inevitably about the war in Europe and the potential of our involvement. There were school dances with the Big Bands—Glenn Miller, Tommy Dorsey, Artie Shaw. Some in the group had cars now and so we had greater mobility. During breaks we took ski trips to New Hampshire, where I first learned of the beauty of winter in the New England mountains—and the exhilaration and hazards of skiing. Skiing was more primitive in those days with rope tows, long clumsy skis, and harnesses of questionable safety. But the crisp air and bright skies and the rush of wind in my face and the ski lodge conviviality were joyous breaks in the scholastic routine.

During my senior year, I made the acquaintance of John Loofbourow, a warm, thoughtful, kindly man in his forties with a real affection for people, who became my mentor. Trained as a physicist in spectroscopy, he had become interested in the effects of radiation, especially ultraviolet radiation, on living cells. He was intrigued by the observation that injury to cells or tissues elicits a response to repair the injury, as with the healing of a cut or bruise. To provide a simpler system for the study of this phenomenon, he had demonstrated that yeast cells, damaged by ultraviolet radiation, released into their medium unknown substances that promoted the growth of other, undamaged yeast cells. John thought these factors, "wound hormones," might be novel substances synthesized by the damaged cells in response to the radiation. He wanted to isolate and characterize them and sought to understand their effects.

Much later it was realized that this result was not a specific response but was primarily a consequence of damage by the radiation to the surface membranes of the yeast cells, greatly increasing their permeability. This caused the cells to leak into the culture medium a variety of vitamins, coenzymes, amino acids, nucleotides, and so on, which accelerated the growth of other yeast in a relatively simple medium.

In addition to his innate kindliness, John had a decidedly pragmatic outlook. Problems were to be solved patiently but with persistence. I hardly ever saw John ruffled or agitated, but nor was he lethargic or insensitive. He became my mentor for much of the next nine years. His wife Dorothea was a physician. They had a teenage son who was already a highly talented pianist. This highly-educated, talented, and urbane family provided a revealing model for me.

John taught me the art of ultraviolet spectroscopy. Nowadays, it is a routine procedure with programmed electronic instruments that can plot out an absorption spectrum within seconds. In those days, to obtain an absorption spectrum involved a tedious photographic procedure. To provide enough light with a near continuum of wavelength, a tungsten spark was used. The light was passed through the specimen, then through a quartz prism to spread out the spectrum, and thence onto a photographic plate. A series of exposures was made with the specimen in the beam and again with a blank in the beam. After development, the plate was scanned with a microdensitometer to find match points at particular wavelengths—regions of equal blackening from sample and blank spectra, taken with different exposure times. From these, the absorption at the selected wavelengths could be determined and the absorption spectrum plotted. It was quite a task to do one complete spectrum in a day. Each newly acquired spectrum was thus a valuable addition to the literature.

The extraordinary improvement of today's equipment—a factor of a thousand in the time required to obtain an absorption spectrum—is an excellent example of the power and role of technology in biological advance.

I learned to perform the "wound hormone" experiment. Growing a sufficient number of yeast cells, irradiating them, incubating the damaged cells for the appropriate period, collecting the culture fluid by centrifugation of the cells, and partially purifying and concentrating the supernatant to a point of stability required a continuous twenty-four-hour stint. More than one night of sleep was lost to this protocol.

Francis Schmitt, the new chairman of the biology program, was a very different sort. Aloof, somewhat Prussian in manner, ill-at-ease and formal with students although basically well-meaning, he did not encourage a personal relationship.

In the summer between my fourth and fifth years I worked as a paid assistant with a small research group at the Michael Reese Hospital in Chicago. It was my first paid job. The group was trying to study an unusual protein they had detected antigenically in the sera of rabbits that had been injected with cells of a transplantable tumor. I learned how to bleed rabbits (although I never enjoyed this), how to observe antigen-antibody interactions, and so on. I undertook preliminary purification of the protein by the limited, but then conventional, techniques. Several were ineffective, but I did have some success with ammonium sulfate fractionation.

I learned how slowly research often proceeds, especially when coupled to the physiological time-scales of animals. I learned how difficult it can be to obtain even modest funds for research support, also the great handicaps placed on a small research effort in an institution not really devoted to research or to teaching but to patient care. It was also educational to note how the physicians were in a different caste, with their own exclusive dining room and other facilities.

A research thesis was part of the fifth-year program. I initially had thought to study the recently crystallized tobacco mosaic virus (TMV). Viruses were still mysterious entities. They could reproduce, but not as free-living organisms. It was thought that they were quasicells lacking some essential component. My idea was to look for known vitaminlike substances in TMV to see if there were specific deficiencies. (We now know that all would have been absent!) However, no one at MIT had any experience with TMV, nor were there the facilities (greenhouses) to grow it in tobacco plants. I therefore set out on a far-too-ambitious program to test John Loofbourow's "wound hormone" hypothesis in a very different setting.

It was known that flatworms (*Planaria*) could regenerate large parts of their bodies when injured. Indeed, if cut laterally, each half could produce a somewhat smaller whole, the rear half developing a new head and the front half a new tail. My experiment was to test whether flatworms injured (as by ultraviolet radiation) would release factors into the medium to facilitate such regeneration. No one at MIT had worked with flatworms, but they could be purchased and, according to the literature, were easy to raise. However, I did not find this so and indeed had continual difficulties throughout the year in persuading the organisms to remain alive and reproduce. They could indeed regenerate as reported. But, at least in my hands, the proportion regenerating and the rates of regeneration were highly variable from batch to batch—so much so that I could not obtain what I could regard as a statistically significant effect by application of the medium recovered from injured flatworms.

Thus, empirically, the research was quite inconclusive. Educationally, I learned much about the need to have a biological system under good control before performing molecular experiments with it. I also learned to appreciate the importance of statistical variation and the need for the ability to reduce such variation in order to detect modest effects. And I learned of the need to have available a source of expertise in the "art" of handling any specific biological system—in this case, flatworms.

Research can be humbling, but fortunately one can learn from mistakes.

John Loofbourow's choice of ultraviolet radiation as the means of injury to the cells had an indirect and ultimately fateful consequence. It was a natural agent for a physicist to use, easily quantitated and varied, but it led me into inquiry as to the nature of the damage done to cells by ultraviolet radiation.

It was known that such irradiation could injure cells, slowing their growth rate even to the point of death. In appropriately designed experiments, ultraviolet radiation had been shown to induce genetic mutations. Measurement of the relative effectiveness of different wavelengths of ultraviolet provided clues to the absorption spectrum of the molecules directly affected by the radiation. Less radiation would be needed at those wavelengths at which these molecules absorbed more strongly.

Such data suggested that a poorly known class of molecules in the cell, the nucleic acids, were the immediate target of the radiation and that photochemical injury to the nucleic acids somehow led to mutation or cell death. Because the cellular functions of the nucleic acids— thought then to be fairly small molecules—were completely obscure, these observations led to a mystery, a puzzle of great significance. But this was to be for the future.

MIT was a serious place indisposed to student revels or frolics, which therefore I missed. But it provided a thorough preparation for a life's work.

Looking back, I see with surprising clarity how the twig was bent;

how the innate talents were fostered, shaped, and channeled by external forces; how, if not the details then the shape and character of my life could have been predicted at age twenty. The pattern was in place— what remained was the unfolding, the evolution and manifold expression of that character.

Science has been a compass in my life, my means of reference to truth and integrity. MIT gave me that compass.

At twenty-two I had now found my calling—that sector of science, the frontier of biology, where I would find a lifetime of intellectual excitement and satisfaction. I was, technically, superbly trained and honed; I had learned some practical lessons in the art of research, in the necessary match between the problem selected and the scale of effort, the facilities, and the skills available. Within science, I had acquired considerable self-confidence and intellectual independence. MIT had shaped my raw talents, girded them with skills, and exposed many paths for their life-long use. Yet socially I was still immature, still enmeshed in the parental ethos. But now my life was to take an abrupt turn.

In 1940, the long-lived isotope of carbon, C^{14}, is discovered by M. D. Kamen and S. Ruben. The availability of C^{14} has enabled biochemists to trace, with precision and high sensitivity, the pathways of carbon atoms and groups of atoms through the synthesis and degradation of the multitude of carbon compounds found in living organisms.

4

World War II—
The Radiation Laboratory

"Radar won the war, the atomic bomb brought the peace" has become a cliché about World War II. In no other war had technology and technological advantage played such a decisive role.

Quite unpredictably, I spent the war years in the role of a radar system engineer-designer and flight tester. But then war produces many distortions, disrupting many lives.

That technology played such a decisive military role was in part a reflection of the increasingly technological basis of our society. In part, it also reflected the happenstance of recent major scientific breakthroughs that could be adapted to important military purposes. To make such adaptations that needed the latest scientific understanding and to do so quickly required all the recently educated manpower the country could muster. My training was in some ways marginal, but I had the requisite background.

My youth was scarred by World War II. All of my MIT experience had been shaded by the growing threat of war in Europe. Each spring and fall brought a new crisis: Anschluss with Austria, Munich, the Sudetenland, the Soviet-German Pact, and finally, in 1939, Poland and full-scale war. Reared in the Midwest, in the isolationist heartland, I largely shared that view. Europe had been a battleground, an arena for senseless, vengeful conflict for centuries. Many of our ancestors, including my own, had come to this country at considerable hardship to escape for themselves and their descendants the burdens of European history with its recurrent cycles of destruction. Why should we now undo their sacrifice to intervene in these savage squabbles?

Europe seemed remote. Over a thousand miles from either coast, separated from Europe and Asia by thousands of miles of ocean, people in the Midwest felt that the conflicts of Europe or Asia were not their concern. Nor did it matter much who won or lost. International trade was small, and distance seemed to provide an invulnerable barrier. No one envisioned the shrunken postwar world of supersonic flight, intercontinental missiles, and nuclear weapons.

Generally, we believed that the United States' entry into World War I had been a mistake; we had salvaged victory for the French and the British, who had then shunted the U.S. aside and imposed a vindictive peace on Germany. We were now seeing the inevitable revenge. Weekly picture sections in the newspapers reminded us of the endless carnage and torment of World War I and reinforced the conviction, "not again."

Looking back at the countless wars of history, I saw conflict as a futile, feckless endeavor, accomplishing little if anything of lasting benefit to humanity. Through the accretion of knowledge, especially scientific knowledge, humankind had discovered a far more enduring means to achieve progress and enhance human life than that of robbing and slaying one's neighbor. After the outbreak of war in Europe, I was relieved that we were not involved and hoped that somehow this neutrality could continue. The prospect of war was not merely one of unpleasantness and danger; it was a complete and useless diversion, a loss of who knew how many years from the scientific career for which I was single-mindedly preparing.

In retrospect, this attitude toward World War II must seem näive. Today, World War II is regarded as the "good war," the necessary war against the forces of a barbaric fascism that sought to destroy much of Western civilization. But during my student years at MIT, I did not read a daily newspaper, I had no radio, and of course there was no television. My days were principally devoted to study and learning. Major news events penetrated our world, of course—the growing prospect of war, the political campaigns. But, other than in very general terms, I was quite ignorant until after the war of Hitler's depravity, of the campaign of genocide against the Jews, and of the systematic slaughter in the death camps.

The bombing of Pearl Harbor came as an utter and stunning shock. Few of us had paid much attention to events in the Pacific. Asia was even more remote than Europe, both physically and culturally. Japanese industry and its military, unlike those of the Germans, were held in low regard. How could they dare to challenge the United States? The extent of the damage was for a time well concealed, but the bombing of Pearl Harbor needed no exclamation point. It was the bell whose toll had long been dreaded. We were now at war—and for how long? The future was suddenly unpredictable, a new fact of life that quickly became chronic.

Only a few days after Pearl Harbor, air raid sirens wailed across MIT. Students, staff, and faculty crowded into the dimly lit basement tunnels that had been designated as the safest places. I recall John Loofbourow and his wife circulating through the tunnels, trying to provide some reassurance, and I remember saying, "Now it comes to this"—huddling in a basement, waiting for destruction. It seemed a depressing step down from the exhilaration of scientific advance.

After the radio announced that enemy planes were fifty miles from the coast, we expected to hear explosions within a few minutes. No one doubted that MIT would be a prime target in Boston, and we had no defenses. As time passed slowly in the tunnels, confusion and doubt began to mix with fear and bravado. After about two hours an all-clear sounded. In truth, the whole episode was a hoax, concocted by the Air Defense Command to test the state of our preparation, but its effect was counterproductive, leading people to disregard subsequent alerts.

Pearl Harbor created a personal dilemma. I was in the fifth year of a five-year program leading to a degree in quantitative biology. My plan had been to continue further graduate study for a Ph.D. The military draft would now make that plan impossible. In addition, after the shock of Pearl Harbor, despite my convictions of the folly of war, I felt a deep moral obligation to contribute to the nation's war effort. Having completed nearly five years of a high-level technical education, it seemed to me that I should in some role be able to make a more meaningful and significant contribution than I might as an ordinary foot soldier. But biology per se—and, certainly not in that era, the basic biochemistry and biophysics in which I was so interested—seemed unlikely to be helpful in this war.

My problem was resolved in a fortuitous manner. My mentor, John Loofbourow, a biophysicist well trained in physics, was asked to join the Radiation Laboratory at MIT, then rapidly expanding. He in turn asked his two graduate students, myself and Roy Slaunwhite, to join the laboratory. Our training in basic physics and mathematics, and more particularly in electric circuits and electronics, would be useful. The very existence of the Radiation Laboratory, started in 1940, had been a secret kept remarkably well from most of the MIT community. We had, of course, seen sheds and strange-looking domes arise on the roof of the Eastman Laboratory, but their function was obscure. Rumors circulated of research into high-frequency radio waves and even into microwaves. But to what purpose? The strange palindrome *radar* was mentioned, but its meaning was unknown. Even the presence, as I later learned, of some very distinguished physicists went unnoticed.

All soon became clear. The Radiation Laboratory, in the spring of 1942, expanded from a few hundred personnel to several thousand. New wooden buildings were rapidly constructed on space made available by tearing down temporary buildings left over from World War I. (The temporary buildings of World War II persisted at MIT until the 1980s.) As we soon learned, the mission at the laboratory was to develop microwave radar. Radar—its name an acronym for radio detection and ranging—was a means to detect distant objects such as airplanes, surface vessels, and surfaced submarines by the reflection of radio waves. By emitting such waves in short, microsecond bursts and measuring the time for the reflection to return, one could, knowing the velocity of radio waves, ascertain the distance of the object. By focusing such waves into a beam like a searchlight's and scanning across a sector, one could determine the location of the reflecting object. The more tightly focused the beam, the more precisely the direction could be ascertained. The shorter microwaves could be focused into a narrower beam with an antenna of portable size that could be fitted onto airplanes.

Today radar is commonplace, used for navigation and control of air and ship traffic, for detection of speeders on highways, even for determining the velocity of baseball pitches. But in 1942, the ability to detect and locate planes and ships at great distances at night or through clouds or fog seemed almost miraculous.

Longer-wave, ground-based radar had played a decisive role in the aerial struggle over England during the Battle of Britain. But to use microwaves for radar required an efficient source of microwave radiation. Such a source, the magnetron, had been invented in England in 1940 and, besieged as they were, the British had brought the magnetron to the United States. The Office of Scientific Research and Development, under Vannevar Bush, had then established the Radiation Laboratory at MIT to exploit this breakthrough. Lee DuBridge was brought in to head the laboratory, and distinguished physicists were recruited from across the country.

Microwaves not only permitted the development of much higher precision radar but also put us one jump ahead of the Germans, who had no such capability. Thus, they at first were quite unaware of our radar surveillance. But because the Germans would in time surely catch up, the laboratory continually pressed on to new and shorter microwave frequencies. Initially, systems were developed using ten-centimeter radiation (S-band); subsequently, three-centimeter radar (X-band) and finally 1.25-centimeter radar (K-band) were developed, the latter of which, unfortunately, turned out to coincide with an absorption band of water vapor in the atmosphere that limited its useful range in some climates. Each new wavelength required the development of new transmitters, radiation carriers (wave guides), receivers, and antennas. Each new application required the development of a system, involving coordinated transmitter, receiver, antenna, and display devices, all adapted to particular functions and vehicles (plane, ship, truck, and so on).

The constant technological ferment of the Radiation Laboratory was illuminating. It gave me a lasting insight into the modern power of sustained, cumulative technological development performed on a massive scale. In the postwar years, a similar process has led progressively to the revolutions in electronics, communications, computers, and space programs. Many years later, this experience prepared me to conceive of and recognize the existence of the technological potential to sequence the human genome.

An essential and novel component of these radar systems, in addition to the use of microwaves, was the use of feedback circuits for self-correcting control, aiming, and so forth. In such circuits, a portion of the system output is "fed back" to be combined with input in such a manner as to converge the output on a desired goal. If the input changes, the system swiftly self-corrects. The theory of the use of these circuits was then new and increasingly elaborated. I was struck by the analogies between such circuits and various processes such as homeostasis in the biological world and all manner of cyclic phenomena in society in general. Biology was always in the back of my mind.

With its urgent mission, the Radiation Laboratory was organized hierarchically and I was a very junior member, to be assigned where needed. I was arbitrarily placed in Division 9, Airborne Radar. Over the next few years, I was thus assigned successively to three different airborne radar programs—a radar for navy night-fighter planes, a tail-warning system for air force fighter planes, and lastly a general purpose navigational radar, primarily for the Troop Carrier Command. I was involved in the design and flight testing of each system and accompanied its demonstration to the military at various airbases. All of this was a continual learning experience of diverse episodes—many boring, some unpleasant, some dangerous, some triumphant, all crowded into a few years that seemed endless at the time.

The X-band radar for navy night-fighter planes, to enable a pilot to locate and shoot down an enemy fighter plane at night or in heavy fog, was already well along in design. An installation of a laboratory-built model was proceeding into a test plane. A hangar had been set aside for Radiation Laboratory use at Boston's Logan Airport and converted into a somewhat makeshift laboratory. Then and there I immediately came to appreciate the gulf between the theoretical understanding of a device and the practical task of making it work. The latter requires not only understanding but well-thought-out procedures, adequate test equipment, and often experience-based knowledge of the likely causes of malfunction. In a novel experimental system, all of these may well be lacking. Ingenuity and intuition are in high demand.

When I first arrived one morning at Logan Airport—we were taken on a bus from MIT each morning and returned each evening—I found my new "chief," Bill Cady, reflecting in some dismay on the disassembled radar system that had just been brought out from MIT to be installed in the test plane. First it had to be assembled on the bench and made operational. When the rotating antenna was installed in an airport window, everything was out of alignment. The entire apparatus—transmitter, transmit-receive box (a device to shield the receiver during the pulse transmission), waveguides, antenna, and receiver—had to be hand-tuned to the magnetron frequency, which in turn had to be chosen to match a resonance in the magnetron cavity so as to maximize the power output.

Because this was one of the very first X-band systems, for test equipment we had only primitive power meters (in certain ranges only a small neon bulb that glowed when exposed to radiation). The process required systematic adjustment, measurement, readjustment, and remeasurement in a complex cyclic process that gradually proceeded through the system until all the components were optimally aligned and coordinated. If the magnetron failed, we repeated the whole procedure.

Nearly fifty years later, I found myself in a similar predicament in seeking to master the operation of a "homemade" version of a very new physical instrument, an atomic force microscope. Lacking appropriate test equipment, the only criterion for alignment was the provision of a clear output, the image. Failure to obtain a clear image could be due to a flaw at any of several stages and alignment was again a delicate reiterative process.

In 1942, the war in Europe and North Africa was going badly. American efforts to resupply Great Britain, Russia, and the Free French were gravely hindered by the German submarine attack. Even along our Atlantic coast shipping was being devastated. The biggest contribution of radar to the war effort at this time was in the detection of submarines by airplanes. The submarines of that era had to surface periodically to recharge their batteries and their hulls then made good radar reflectors. Homing in on these, and undetected because of the use of microwaves, the radar-bearing aircraft were able to substantially reduce the submarine toll on Allied shipping on our East Coast and in the Atlantic.

Spurred on by this example, we felt a real sense of urgency—the military desperately needed this new equipment. Sometimes the contrast between our tasks of performing abstract calculations and machining parts and that of a navy pilot straining in the dark to see an attacking enemy plane seemed bizarre. But such is modern warfare.

We worked Saturdays and often Sundays into the evenings. Finally, the equipment was installed in a two-seater fighter and test-flown. I was then assigned to accompany the plane to the Quonset naval base at Narangansett, to keep the equipment operational while navy pilots conducted tests on simulated missions.

Seldom have I felt so out of place. Life for a civilian at the base was generally unpleasant. I was assigned a room in the Bachelor Officers' Quarters (BOQ). Most personnel were confused as to what a civilian was doing on the base, and I was regarded with understandable suspicion. Why was this civilian wandering about the hangar and into high-security areas? I had no permit to go off base and no transportation to do so anyway.

Evenings were spent with the pilots who were flight-testing the equipment. We had little in common—young, cocky, and seemingly eager to get into combat, they were far from academic types. Their evenings were spent in heavy drinking and swapping endless flying stories. Apparently, all junior officers were expected at some time in their stay at Quonset to fly their planes in forbidden patterns under the Naragansett Bridge as evidence of their skill and daring. For this they would be nominally reprimanded by the commanding officer and toasted with champagne at the Officers' Club.

On weekends the pilots all received passes and cleared out for Providence, Newport, or New York. I was left more or less stranded.

Fortunately (for me), after two weeks a major accident occurred. A portion of the rotating antenna support cracked and the spinning antenna broke loose, damaging not only itself but the radome (plastic covering) as well. This was beyond my capacity to repair, so the plane and I returned to Boston. The navy at that point decided that no further testing was necessary. The equipment was taken out of our hands and put into production and later used successfully.

I had intended to entertain no thought of marriage until I completed graduate study, but the war promised to postpone that time indefinitely. Most of the Radiation Laboratory personnel had families; the laboratory did not provide the comradeship of classmates. I had become close to Joan Hirsch, from Chicago, then in nurse's training, whom I had met a few summers earlier. In August 1943, we were married in Chicago. After a very brief honeymoon in New Hampshire, we found a small apartment in Cambridge and Joan soon found a job as a laboratory technician at Boston City Hospital. Our marriage produced two daughters and a son. Some twenty-eight years later, however, our interests having greatly diverged, it ended in divorce.

After the night-fighter project, I was assigned to a small project to explore the possibility of an automatic radar tail-warning device for air force fighter planes. This design brought me in contact with the diversity and wealth of talents and skills that had been assembled.

The Radiation Laboratory was a remarkable place. It was perhaps the first laboratory to undertake research and development on such a grand scale. Also extraordinary was the virtually unlimited availability of resources. The stockroom was free! And the stock was swiftly replenished by knowledgeable purchasing agents. One simply went to it and took whatever resistors, capacitors, vacuum tubes, and so on one needed. At MIT, in the chemistry stockrooms, every item, every test tube, beaker, and bunsen burner had been charged against one's account and returned for credit if still in "good" condition. The difference in paperwork was revealing.

Expert help on any aspect of radar theory and technology could be readily obtained. Specialists were continually seeking to improve transmitters, receivers, antennas, display devices, and so on. The laboratory had brought together an extraordinary group, mostly young physicists, who were applying their theoretical knowledge and their native ingenuity to a single multifaceted goal—advancement of microwave radar. Within the laboratory, knowledge was, with a few security exceptions, shared freely. Weekly evening seminars were held to disseminate the more recent advances. Project reports (classified) were written and available in the laboratory library. A few projects were, for unclear reasons, regarded as more highly classified, but the knowledge of their existence and parameters inevitably spread within the community.

All the work at the Radiation Laboratory was under security clearance, and a photo badge was needed for admission. Shortly after Pearl Harbor, a military guard was placed at the laboratory. Secrecy was taken seriously—I did not discuss the nature of my work outside the laboratory, even in my family. This created an unfortunate habit of reticence, which in part has persisted long after the reason for it disappeared.

The military presence guarding the laboratory persisted until 1943 when an associate director was shot while leaving the laboratory, one evening, for failing to halt. Less security was then deemed sufficient.

In 1943, the Eighth Air Force began its major bomb attacks on Germany. A key laboratory project at this time was the construction of a dozen advanced X-band terrain-scanning radars to be installed in the Eighth Air Force bombers in England. These were to be used as lead planes to guide bombing missions over Germany in the frequent bad weather or at night. A number of laboratory personnel accompanied these to England to oversee their installation, maintain the equipment, and suggest modifications in use.

Outside the laboratory, life was becoming increasingly unpleasant for everyone, albeit much less so than for those in combat. Increasingly stringent food rationing restricted our diet. We had no car and could not have found gas if we did. All transportation was overwhelmed with troop movements, and any travel became arduous. When my father-in-law died in Chicago, the trip to the funeral required two twenty-four-hour stints in overcrowded train coaches that gave me a severe bronchitis.

As draft registration had begun while I was a student, I was registered in my home district of Chicago. The draft board there, understandably, regularly sought to draft me. Each time, the laboratory filed an appeal on grounds of the national importance of our work and, each time, the appeal was granted whether at the state or national level. However, only a six-month deferment was granted, so every six months the process started over again. As the appeal often took several weeks, I was repeatedly required to take the standard physical exam for draftees in Boston and on one occasion even received orders for induction before the appeal was finally granted. This apprehension—the recurrent threat of imminent major upheaval—was disconcerting.

Actually, it later became evident that it was not possible to leave the laboratory even if one desired. A few of the younger scientists decided they would rather serve in the military and signed up. Imagine their surprise when they were immediately assigned back to the laboratory, only now in uniform.

The tail-warning system was a technical gamble from the beginning. Given the weight and power limits that fighter planes could afford, quick calculations suggested that the ability to detect an enemy plane at a useful range would be marginal. Nevertheless, the importance of the problem suggested that it was worth a design and feasibility test and assembly of a prototype. Bob Taylor, a scientist of similar age, and I were assigned this project.

The only hope of providing a radar reflection of sufficient strength to be unambiguously and automatically detectable seemed to lie in an integrative approach that would rely on the repetitive character of the returning echo contrasted to the sporadic character of the background circuit "noise." Because, however, the pursuing plane would be closing in, the time available for integration of successive echoes was strictly limited. With today's computers and electronics, this would be a far more feasible task. Using the electronic techniques then available, we were in fact able to detect quite weak signals and use these to activate a bell or light for the pilot in the cockpit. We then assembled a prototype that (barely) fit within the weight limits and installed it in a Cessna aircraft for flight testing.

Testing was somewhat hampered because it always required two planes, our own and a target plane. We also soon learned that the pilots had very limited ability to judge the distance of a second plane. So we had to modify our equipment, which had no readout other than a warning signal, with a test monitor to provide range information. We also had to automatically limit the range of detection to a distance less than the plane's altitude or else the radar reflection from the ground would trigger the warning.

We soon discovered yet another problem. It had long been known that some clouds, especially cumulus clouds involved in thunderstorms that accumulated electric charge, could give weak, diffuse radar echoes. Indeed, this property is now used in meteorology. On a radar screen this caused little problem; however, with our increased sensitivity and our designed ability to integrate echoes over a wide sector, we found the thunderheads were not infrequently setting off our alarm. When we reported these results, it was generally agreed that the project, if feasible at all, would be more likely to succeed at longer, submicrowave frequencies (for technical reasons having to do with antenna effectiveness and lesser reflection of clouds). The problem, in any case, did not require the narrow beams for which microwaves provide a significant advantage. The project was thus terminated and I was assigned to a new one then being initiated, the design of a general purpose terrain radar, initially intended for use by the Troop Carrier Command in locating their drop zones.

This experience illustrated a negative aspect of work at a junior level within a large laboratory. One may be assigned to a problem and work on it with great diligence, energy, and imagination only to be informed that one's solution will not be implemented. In basic science, the goal is simply knowledge. In applied science, the goal is practicality as measured by military, economic, or political criteria.

The years of war dragged on. Coping with the daily minutia of system design, airplane installation, and test flights repeatedly postponed because of weather or aircraft maintenance delays, it was difficult to sustain the earlier sense of urgency. The work was challenging, requiring improvisation and ingenuity, but now connected through some lengthy channel of time to the military struggle. And my real interests were elsewhere.

By now, early 1944, the invasion of Europe was expected within the year. We were more knowledgeable of the time and scale of effort required to design, test, and shepherd into production each new radar system. Was it sensible to launch entirely new projects? The response was that Washington expected the war might continue as long as ten years. Success in Europe might require two to three more years, to be followed by, it was expected, a battle of near extermination against the Japanese army, which would never surrender. It was a grim prospect. While my situation was surely better than that of those in combat, it appeared that my entire youth would be spent in an unwanted war, doing development research in a field of only secondary interest to me.

The ability of radar to distinguish features of terrain was evident early on. Water was a poor reflector, but coastlines and the shores of lakes were readily detected. Structures, too, and clusters of structures were good reflectors, though at low altitudes hills and ridges created obscuring radar shadows. The finer beams of three-centimeter radar permitted more precise resolution of details.

The immediate aim of the new project, dubbed APS-10, was to use the most advanced techniques available to develop a sophisticated, lightweight, relatively low-power radar to be used by the military for general navigation and by the DC3s of the Troop Carrier Command for specific troop drops at night by following previously established radar "maps." The APS-10 was also conceived to be a prototype for possible postwar radar for civilian aviation.

Design and assembly of the system went relatively smoothly. I was familiar with every aspect of the system and was subsequently given responsibility for installation of the first test system into a DC3. The Radiation Laboratory had outgrown its facility at Logan Airport and was now using Bedford Airport near Lincoln, Massachusetts. It was a longer busride and there was no alternative means of transport should I miss the bus.

DC3s were manifestly not designed to accommodate radar equipment. To provide a radar plot of forward terrain, the scanning antenna had to be located in a plastic dome beneath the fuselage. Aerodynamic consultants indicated that a dome of the size needed would *probably* not threaten the craft's airworthiness.

By May 1944, installation was at last completed, the equipment was operational, and all was ready for our first test flight. When we arrived at the airport, however, we learned that disaster had struck. Our air crew had without authorization taken our plane that morning to an airbase in New Hampshire to pick up some parts needed for maintenance of another plane. Unfortunately, visibility was poor at the New Hampshire base, and our plane had somehow struck the side of a mountain, killing the crew, disintegrating the plane, and demolishing our radar installation. This shock reminded us of the hazards of our occupation.

But flight testing of the radar design was still required. Assembly of another radar system would take but a few weeks. Acquisition of another DC3, its modification with a radome, and installation of the radar would need three months or more. We therefore made an impromptu installation of the radar in the nose cone of an A25 fighter-bomber. Normally, this site was occupied by the gunner. We crammed in the radar antenna, transmitter, receiver, and display unit. As the operator, I had to sit in the nose cone just behind the antenna. Remarkably, this all worked surprisingly well. While regulations prohibited anyone to be in the nose cone during takeoff and landing, the equipment made getting in and out during flight much too onerous a task. The landings were quite exciting with only the dome of clear plastic between myself and the runway a few feet away.

When the second DC3 installation was complete and operational, we took the equipment to the military for scrutiny by its personnel. Our first assignment, in November 1944, was to the Troop Carrier Command in Indianapolis. Shortly after arriving there, we discovered that a crucial piece of test equipment needed to keep the system operational at near peak performance was not available. When I was sent back to Boston to bring one out a bizarre incident resulted.

On my return, an air force general who had been visiting the Radiation Laboratory was flying back to Wright Field in Dayton, Ohio, and it was arranged that I would carry the equipment on board his personal plane. We arrived at Wright Field late in the evening, and I was put up for the night at the BOQ. My equipment was left on the plane. Next morning, I sought to retrieve the equipment to take it by train from Dayton to Indianapolis. Once again, I was an odd civilian poking about high-security hangars. I was quite certain as to which hangar held the plane in which I had arrived. To my dismay, it had been moved. When I inquired, no one had any knowledge of the plane. Remarkably, there seemed to be no record of its arrival.

It was clear that my inquiries were breeding dark suspicions of my interest and perhaps my sanity. Finally, on the verge of being turned over to the MPs, I persuaded an unbelieving sergeant to call the general's office. With some difficulty, I was able to reach the general, who with evident annoyance was able to resolve the situation and arrange for the return of my equipment. I thereupon as quickly and boldly as possible strode out through the airfield gate, suitcase in one hand, equipment in the other, to catch a bus to town. That I was able to simply walk off the military field with a piece of classified electronic equipment with no pass or identification still mystifies me.

This incident notwithstanding, secret war research on applications of high technology to defeat a sophisticated foe was not as glamorous as one might expect. The reality, at least in our roles, far removed from the levels of decision, was often tedium.

In Indianapolis, the weather was, as might be expected in November, miserable. We could fly perhaps every third day. On the days we could fly, we went up two or three times at troop carrier drop altitudes of about five hundred feet to avoid enemy radar. At such altitudes, the turbulence was constant—every gust, every thermal, jolted the plane. It was a good test for the equipment, which performed well, and for us who did not perform as well. I became airsick nearly every flight but "carried on." After about three weeks, the Troop Carrier Command seemed well satisfied with the equipment and when the weather cleared, we returned to Boston. This was my first extended night flight and I marveled at the lights of the cities and towns against the deep blackness of the countryside. Greater New York was a vast sea of light.

Our next demonstration, to be held in Orlando, Florida, seemed more promising. We took off from Bedford early one frigid January morning (it was minus ten degrees) and, after the serious mishap en route, described in the Introduction, arrived at balmy Eglin Field in Orlando. As a northerner, I had frequently read about or seen pictures of Florida in the wintertime, but the reality of this sunny, lush land less than a day away from the frozen gloom of New England was startling. Today, with television and jet travel we are accustomed to these contrasting realities. In 1945, this was a novel, almost bewildering experience. At the same time, I was shocked and offended by my first exposure to the segregated restrooms and drinking fountains of the South. And these were, of course, but the most superficial evidences of racial discrimination.

Compared to those in Indianapolis, demonstration flights for the air force at Orlando were easy. We flew over sea to the Bahamas and over land in Florida. The equipment performed well and the air force personnel seemed pleased. Except for the exceptional size of the cockroaches, life in Orlando seemed idyllic. After two and one half weeks we returned uneventfully to Boston.

Next, the APS-10 had to be put into production. The General Electric plant in Schenectady, New York, had in some way been selected as the manufacturer and I was sent to there as liaison to the engineers. Here was yet another unfamiliar and hectic milieu. As I soon learned, the problems of mass production of a few thousand radar sets, plus spare parts and technical manuals, to be operated by military personnel, are very different from the construction of a few prototypes to be operated by highly trained scientists and engineers. Cost factors, simplicity of controls, ease of maintenance and repair, adaptation of design to production-line manufacture, even aesthetics of appearance, all became important elements that could not, however, be allowed to compromise performance. I am sure the GE engineers were very good at their jobs—as an academic researcher, however, I was not always fully sympathetic with their problems or solutions.

Schenectady was not so far from Boston, so I shuttled back and forth. Thus, I was at the Radiation Laboratory on VE Day in May 1945. It was a joyous celebration, taking off from work, at least informally. A group of us gathered at Art Solomon's house to hail the victory. Many whose interests were more focused on Europe seemed to feel the war was over. Others of us still thought of Japan and the blood yet to be shed. Nevertheless, from that day on, we could begin to foresee the end of the Radiation Laboratory and to envision the return to a postwar world.

Soon, we learned that the long-held secret of the Radiation Laboratory was at last to be told. Our activities and the important role of radar in the war effort were to be made public. At last I could let my wife and family know what I'd been doing for the past three years, although it did indeed seem strange to have (selected) reporters prowling about the laboratory, asking questions, taking pictures, and so on. The release date for the story was, for obscure reasons, repeatedly postponed. It was finally set for the first week of August—as it turned out, the day of the bombing of Hiroshima.

Rumors of some kind of atomic development had spread from time to time around the laboratory. Several of the senior people—Luis Alvarez, Robert Bacher, and others—had rather abruptly disappeared. When asked, we were simply told they had "gone West." I was aware of some large projects in Tennessee and in Washington State. I was unaware of Los Alamos, and my imagination conceived only of an atomic-powered engine that might power an airplane or a submarine for unlimited range—not a bomb!

Thus, Truman's announcement on 6 August was a stunning surprise, although I immediately appreciated its epochal significance. Nagasaki, on 9 August, answered the question of whether we had more than one bomb, and VJ Day truly ended the war.

As we now know, many of the scientists who worked on the development of the atomic bomb anguished over the invention of such a devastating weapon of mass destruction. I would today have similar concerns about the development of biological weapons.

Since Vietnam, many scientists have questioned the morality and ethics of the application of science to military pursuits, but during World War II I had no such qualms. We had been wantonly attacked.

(While there are revisionist historians that maintain that President Roosevelt's policies invited such attack, I believe this to be a misreading of history. The America I knew did not want war.) We had no choice but to defend ourselves. If the application of our science and technology could shorten and help to win the war, save lives and punish aggressors, and thereby perhaps deter future aggression, then we should surely use our talents toward that end.

Radar is, to be sure, a relatively passive and defensive technology that provides information. In war, radar per se never killed anyone. But one cannot escape the fact that in war the lines between offensive and defensive are constantly blurred. The radar that guided bombers to German cities at night certainly had a part in the resultant deaths. I had no qualms, because the end—a just end—necessitated the deplorable means. Hiroshima and Nagasaki frightened me as a portent of the future, but they did not revolt me. From all that I knew and know, they spared hundreds of thousands or more American casualties. And that was enough. We cannot always choose the world in which we live.

I know that for some who survived, World War II was the greatest adventure, the most vibrant time of their lives. Not for me. I've always felt, in a sense, cheated by World War II. By luck, I did not have to endure the agony and danger of combat, but four of the potentially most creative, most carefree years of my life—irretrievable years of greatest imagination, of highest energy, of most recent training—four years of intense and draining effort had to be spent in an alien pursuit.

Detoured from a nascent career in biophysics, my reactions to the Radiation Laboratory were atypical. Most of the scientists were physicists, Ph.D.'s of varying degrees of professional maturity. For them, the Radiation Laboratory was an opportunity to advance, at a very accelerated rate, a sector of their science, as well as an opportunity to utilize to the fullest their training and skills in a project of great importance to the war effort. While recognizing the latter, I, in a much junior role, could utilize only a portion of my training and aptitudes.

My entire generation had been forced to wage a war we never sought. Never again could I know the simple, secure Midwestern insularity of my youth. The war was an epic of courage and national will, but it also forever destroyed the youthful illusion that one had ultimate control over one's individual destiny.

In 1944, in the midst of the carnage of World War II, Avery, McCarty, and McLeod at the Rockefeller Institute demonstrate that the "transforming principle"—a substance extracted from one strain of Pneumococcus that could, hereditarily, alter the properties of a second strain, so as to make it resemble the first—is, in fact, composed of DNA. This most surprising and significant result provided the first direct evidence that DNA played a major role in the transmission of hereditary information.

5

Transition 2

The war was finally over.

The future was now open, but which path to choose? I had been away from biology for four years. I had of necessity acquired a considerable and valuable expertise in some of the most advanced areas of electronics. This skill was soon recognized. Major electronics firms—RCA, General Electric, Bell Laboratories—were gearing up for the return to a civilian economy. Realizing the considerable pool of talent at the Radiation Laboratory, recruiting teams soon appeared. Also, new electronics companies—some small and others not so small—were being organized, several by the new Radiation Laboratory alumni. I was interviewed by several of these companies and offered employment at salaries approaching ten thousand dollars per year, a quite handsome sum in those days and three to four times my Radiation Laboratory salary.

My alternative was to return to graduate school to obtain a Ph.D. But how would I and my wife be supported? It seemed unreasonable to ask my family to provide my support again, since I was approaching the age of twenty-six. For the interim, the Radiation Laboratory had undertaken to provide a written record of its accomplishments and thereby make publically available all of the technical knowledge the laboratory had collectively acquired. The Radiation Laboratory Series ultimately came to twenty-eight volumes. I agreed to stay on a few months to provide two chapters: one on the conception and design features of the APS-10 radar, one on radar altimeters.

The money from industry certainly was tempting and to discard four years of hard-learned expertise was grievous. But in my heart I was a biologist. I wanted to explore the mysteries of living organisms, not to design sophisticated electronic gadgets. Manifestly, a life in the electronics and communications industries over the past four decades would have been exciting and surely more financially rewarding, but I have never regretted my choice.

I became aware, through Art Solomon, of a proposal for the American Cancer Society to provide fellowships to make possible the entry, or re-entry, of several of the younger scientists at the Radiation Laboratory into biology. I applied for this program; others included Bob Taylor and Ed McNichol (later director of the National Eye Institute). Approval of the proposal and award of a fellowship of two thousand dollars followed. Where should I study? I briefly considered other institutions, including Caltech. But we had a place to live, Joan had a job, and my mentor, John Loofbourow, was returning to MIT. There seemed no especial advantage to going elsewhere and several disadvantages, so in February 1946 I enrolled as a graduate student in biology at MIT.

6

Graduate School

The hour could not be over. I had only completed two of the four test problems. And this was my first examination in graduate school.

After Hiroshima and Nagasaki, it seemed clear that nuclear physics—isotopes, radiation—would be an important aspect of biophysics, so I had enrolled in Robley Evans's course in introductory nuclear physics. I went to class, read the assignments, understood the subject matter, and thought I was doing well until the first test. Disaster. I knew how to do the problems in principle. But in four years of pragmatic experiment and development, my advanced mathematical skills had rusted. I barely completed half of the test and received a deserved D.

This abrupt failure (a D grade is failing in graduate school) singularly shocked my self-confidence. Had I been out of school too long to come back? Had I made the wrong choice and taken an impossible path? The problem-solving approach came to my aid. Manifestly, I had to review and revive my calculus, differential equations, vector analysis, and so on. And quickly. I also decided I should know more advanced mathematics and subsequently took a year of advanced calculus. By the end of Robley Evans's course, I was doing much better.

Returning to the academic schedule as a student, after four years' absence was not an easy task. The routine of fifty-minute classes, nightly assignments, and reading initially seemed very confining and required considerable self-discipline.

I was not alone. MIT was besieged by returning students. Courses and classes ran year-round and leisurely summer vacation periods were totally discarded. The Boston climate was unchanged, however; our small apartment opened only onto a narrow court, so some stifling summer nights we slept on the building roof for a bit of breeze. Air conditioning was still in the future.

One of the casualties of the war was the network of friends established during my undergraduate years. While graduation would have frayed the network in any case, the war dispersed my set of friends in all directions and made it impossible even to maintain contact. Now, after four years, the strands were severed and the relationships lost. As if in compensation, our apartment building now housed several couples of returning students, enrolled in graduate programs at MIT or Harvard, among whom a natural camaraderie soon developed. It was a diverse and interesting group of future scientists, engineers, lawyers, physicians, historians, literary scholars, and the like.

The MIT Biology Department had changed considerably in personnel. Frank Schmitt had filled out his faculty, including Dick Bear who did X-ray crystallographic studies of fibers (actin, tropomyosin), electron microscopist Cecil Hall, a surface chemist named David Waugh who was interested in cellular membranes, and Stanley Bennett, an M.D. interested in embryology, among others. I enrolled in the new biology graduate courses—embryology, genetics, biophysics—and started on a thesis under John Loofbourow.

Graduate school is normally a time of specialization and apprenticeship—of learning, under a mentor, the arts and standards of research in preparation for one's future independent career. One's mentor usually proposes one or more research problems in the field of his interest for which he can provide background knowledge, technical expertise, and the necessary facilities and equipment. The wise mentor will also select problems that can reasonably be resolved within a normal period of graduate study.

My graduate experience was, however, anomalous. Four years at the Radiation Laboratory had taught me the ways and vagaries of experimental research, albeit in a very different mode and with extraordinary support and resources. And so, while adapting to a different field, I was already sufficiently prepared and confident to engage in novel, relatively independent research projects. In this mode, I came to know John Loofbourow much more intimately. Thoughtful and pragmatic, he also had a deep compassion for people and empathy with their concerns. Frank Schmitt made John the executive officer for biology as he was much better able to cope with the personalities of faculty and students.

John's "people skills" similarly led him to become active in the MIT faculty senate and he was chair of the faculty in 1949–50. Unfortunately, all of this valuable citizenship took him away from his science.

He loved to tinker, and his pride and joy and diversion was a lathe at which he would turn out often ingenious "gadgets" for laboratory use. He had a long-standing interest (from boyhood?) in the South Pacific and regularly subscribed to the *Pacific Island Quarterly,* a gossipy journal devoted to the economics and politics and personal idiosyncrasies of planters and other characters among the South Pacific islands. In "bull sessions," we would occasionally rhapsodize about taking over an island in the South Seas and, using modern methods, developing its agricultural and mineral resources. A form of Shangri-la.

I resumed my interest in things ultraviolet. This resulted in my first published paper and, more important, naturally extended to nucleic acids as one of the principal ultraviolet absorbing components of cells. A dramatic experiment conducted during the war by Avery, McLeod, and McCarty at Rockefeller Institute had suggested a most significant role for these previously obscure substances. They had shown convincingly that the "transforming principle," an extract of some strains of pneumococcus that had the property of transforming certain characteristics of other strains of pneumococcus in a stable, *inheritable* fashion, was composed of deoxyribonucleic acid. This result, consistent with earlier indirect evidence, strongly implicated DNA as a genetic factor, at least in these organisms.

This result was a solid fact in a bog of speculation. The proteins, it was clear, were the catalysts, the machines of the cell and the body. But whence came the proteins? What created these complex molecules? How did some cells in the pancreas know how to make pepsinogen and trypsinogen (digestive enzymes) and other cells in the pancreas know how to make insulin? This "know-how" clearly was inherited from generation to generation—but not, it seemed certain, by the simple passage of (say) insulin molecules to serve as models. What then was inherited? The concepts of information theory were still nascent in the fields of electronics and communications and their application to biology was yet to emerge.

A few scientists had pondered this problem qualitatively and speculatively. Max Delbrück, reflecting on the faithful replication of genes (whatever they were) over many generations, had wondered whether new, unknown physical laws might be involved. Erwin Schrödinger, in his 1944 book, *What Is Life?*, followed Delbrück in his quest for a mechanistic explanation for genetic phenomena but achieved only the somewhat paradoxical notion of an "aperiodic crystal." (It is of historical interest that Friedrich Miescher, who had discovered the nucleic acids in 1869, proposed in the 1890s that the templates for inheritance might somehow be provided by an appropriate string of asymmetric carbon atoms. The chirality [handedness] of tetravalent carbon had only recently been elucidated.)

The Avery experiment pointed to DNA as the substance of the genes. Recent experiments by Beadle and Tatum at Stanford had demonstrated that the genes specified not only such complex traits as wing shape and bristle pattern in drosophila but much more discrete characteristics such as the presence or absence of one or another enzyme needed to make essential cell components. Genes, enzymes, proteins—an amorphous pattern of hints and glimmers was emerging. The nucleic acids, as yet so ill-defined, could be the key.

My first published paper, appearing in 1947, was only a short note in *Nature,* but I felt a deep satisfaction that it had been accepted. Now I had joined the ranks of those who had made some enduring contribution to human knowledge, my work was part of the stream (today one would say flood) of scientific publication. It was certainly not a major, nor I fear particularly useful, contribution but it was novel. It illustrated well my capacity, at that stage of my career, to define and resolve an unsolved technical problem by bringing together a variety of information and skills.

Science is frequently said to be a young person's game by those who point to the relative youth of those scientists who most often make the greatest discoveries. The reason for this is that science is the art of the feasible or, at the frontiers, the barely feasible, for the readily feasible most often has been done. A young scientist quickly learns what experiments are feasible, barely feasible, or infeasible, and these last are usually mentally discarded and removed from consideration. With time and invention, such experiments may indeed become feasible later. But the barrier against even considering them often remains strong in the older scientist's mind, who thus appears—and is—less creative.

This need, especially in periods of swift advance, to unlearn what "cannot be done" accounts in large part for the advantage of freshly trained minds.

The idea was to make a simple but adjustable filter to isolate various portions of the ultraviolet spectrum for use in studies of the effect of ultraviolet radiation on biological objects. Monochromators, using quartz optics and prisms or gratings, could accomplish this but were expensive and generally required the use of narrow slits at entrance and exit that restricted the energy available and the size of the area to be irradiated.

Christiansen had described a set of very simple filters for the visible region of the spectrum. These consisted of small rough chips of a transparent glass immersed in an appropriate transparent liquid, all contained in a transparent cell. The liquid was so chosen that its refractive index matched that of the glass at a specific wavelength. At this wavelength, the cell containing the chips and liquid was completely transparent to a beam of light. At other wavelengths, the refractive index of the chips and liquid would differ and the light would be reflected and scattered out of the beam at each chip-liquid interface. The further the wavelength from the matching wavelength, the greater the scattering and the lower the transmission of the cell. In principle, the cell could be of any size.

While such Christiansen filters had been made for all visible wavelengths, none had ever been designed for the ultraviolet region of the spectrum. I set out to make one.

First, I required chips of a material transparent to ultraviolet. Second, I needed a liquid, transparent to ultraviolet, with a refractive index to match that of the chips only at a desired ultraviolet wavelength. I preferred to use a mixture of two liquids so that, by varying the composition of the mixture, I could change the selected wavelength of complete transparency. Data on refractive indices of liquids and crystals at ultraviolet wavelengths was scarce, but by literature research and extrapolation, I was able to eliminate most combinations. Only saturated hydrocarbons would be expected to be transparent in the ultraviolet regions and the larger molecules would have the higher refractive indices.

It finally appeared that ground fluorite, which was ultraviolet transparent, could be used for the chips and that its refractive index could probably be matched at ultraviolet wavelengths by an appropriate mixture of the hydrocarbons cyclohexane and decalin (decahydronapthalene). When purchased, the cyclohexane was transparent to ultraviolet but the decalin was strongly ultraviolet-absorbing. No ultraviolet absorption spectrum for decalin had been published, but the measured absorption seemed implausible to me. I decided it must be due to unsaturated hydrocarbon impurities and undertook to purify the decalin by passage through a column of activated silica gel, which I believed would preferentially absorb unsaturated molecules.

The column worked. The eluate was ultraviolet transparent. When mixed with cyclohexane, combinations were obtained that, when added to a cell with fluorite chips, produced the desired filtering effect at chosen wavelengths. I had solved the problem.

Unfortunately, in practice the filters are not very useful. The drawbacks are that the wavelength bands emergent from the filter are quite broad (about 10 nanometers at half-maximum intensity) and, more seriously, that the filtering action is *never* complete. Transmission is still 1 percent or so at wavelengths far removed from the maximum. Since the biological effectiveness of different ultraviolet wavelengths can vary by more than a hundredfold in some applications, such a degree of wavelength contamination is often unacceptable.

John Loofbourow's long-standing interest in the ultraviolet absorption of biological substances was revitalized by the wartime work of Caspersson in Sweden, who had combined ultraviolet microscopy with spectroscopy to determine the ultraviolet absorption spectra of specific cellular structures. Technically, this work was, even by 1940s standards, rudimentary. The optics, dating back to Köhler and von Rohr in 1904, were monochromatic (uncorrected for wavelength dispersion), the photographic measurements tedious and imprecise. However, the potential utility of the technique suggested that it would be worth a significant effort to improve the instrumentation and then apply an improved instrument to a variety of biological questions.

John gathered a diverse group of people around him for this project. Paul Lee undertook the design of a reflecting microscope that would be inherently achromatic. I undertook to design an electronic, self-compensated photoelectric system for measurements. Jesse Scott, an M.D. who had joined us, undertook to develop techniques of specimen preparation for observation. Because I was accustomed to the facilities and services of the Radiation Laboratory, this project seemed to proceed very slowly on all fronts—which was just as well, as I had to prepare for the dreaded Ph.D. examinations.

These examinations are both written and oral and are expected to be comprehensive and challenging. As the first postwar crop of four candidates in a virtually new department in which precedent was lacking, we students were quite unsure of what to expect; I realized later that the faculty was similarly unsure of what level of performance could reasonably be anticipated. The resultant examinations were in fact very difficult and I was the only one of the four to pass without condition.

Development of the ultraviolet microscope continued to be delayed by the time required for design and fabrication of the mirror optics. (Today's computers would have greatly accelerated this work.) My attention was drawn to some papers demonstrating the sharpening of the ultraviolet absorption spectra of benzene and similar aromatic molecules, as vapors, at very low temperatures ($-190°C$, the boiling point of liquid nitrogen) with the attendant revelation of fine detail in the spectrum. Would the same be true of the purines and pyrimidines of DNA? Could the utility of absorption spectroscopy for biological research, at either the macro or micro level, be enhanced at very low temperatures? Would sharpening permit greater discrimination among the absorbers of ultraviolet irradiation within a cell? (Simple cells could be quickly frozen, irradiated, and revived.)

Jesse Scott and I undertook such experiments, initially with purines and pyrimidines. We had first to find a way to pass ultraviolet light through such molecules at low temperature. They would not vaporize at low temperature; simply freezing aqueous solutions would result in an opaque polycrystalline mass. We developed two methods. One involved solution in a mixture of ether, isopentane, and alcohol (EPA), which would dissolve purines and pyrimidines and which froze to a transparent glass at liquid nitrogen temperature. As an alternative method, we learned how to sublime purines and pyrimidines onto a quartz slide by moderate heating in a high vacuum. The absorption spectrum of the resultant thin film was readily measured.

We had made a quartz Dewar vacuum flask with plane entrance and exit windows for transmission of ultraviolet light. After filling it with liquid nitrogen, we could suspend, into the ultraviolet beam, either the EPA solution in a quartz cell with plane windows or the thin films of purines or pyrimidines on the quartz slides. Very appreciable sharpening of the absorption spectra was obtained.

Unfortunately, neither of the techniques was applicable to nucleotides or to DNA. Our only recourse was to dry down samples onto quartz slides. These absorption spectra regrettably showed little effect of low temperature. (We prepared our own DNA, directly from calf thymus glands. The DNA available commercially at that time was often badly degraded. Much elegant physical chemistry performed on DNA in those early years was worthless because the DNA preparations employed were so degraded.)

It seemed desirable to go to still lower temperatures to look for an effect. At that time, liquid hydrogen (at $-252°C$) was available at MIT, but in another building, several hundred yards away. We had to resort to a double vacuum Dewar—the outer Dewar to be filled with liquid nitrogen to insulate the inner Dewar, which would hold the liquid hydrogen. Both Dewars had plane quartz entrance and exit windows so that, when all was aligned, an ultraviolet beam could be sent through the apparatus. The equipment was mounted on a wooden board. Both Dewars were filled with liquid nitrogen. I then carried the apparatus by hand to the source of liquid hydrogen, displaced the nitrogen in the inner Dewar with hydrogen, and returned to our laboratory to insert the specimens and measure their absorption.

We had several successful runs, but the lower temperature did not produce appreciably sharper spectra than we had previously obtained. A disastrous accident (described in the Introduction) subsequently terminated this research.

Because of the constraints of this apparatus, relatively small beams of fairly high intensity were passed through the specimens. In some instances, examination of the specimen after the experiment suggested irradiation damage. This recalled the much earlier observations that ultraviolet irradiation of cells, most likely affecting the nucleic acids, resulted in cell injury and genetic mutation. What was the chemical basis for these effects? What did ultraviolet irradiation do to nucleic acids? I thought it might be possible using our low temperature techniques to obtain clues to at least the nature of the primary radiation effect.

I began with irradiation of the simpler pyrimidines and purines in aqueous solution and quickly obtained a surprising result. On irradiation of the pyrimidine uracil, its ultraviolet absorption largely disappeared then very slowly reverted to the original spectrum. This reversal could be greatly accelerated if the solution were made acid. By the spectroscopic criteria, this spontaneous reversal recovered, almost quantitatively, the original compound. At just this time, Albert Kelner described the phenomenon of photoreactivation of ultraviolet damage in bacteria. We naturally speculated that our reversal phenomenon could be related to his biological reactivation. (Actually, there is no simple relation.)

It was time to leave the scholastic nest. I had passed my examinations and had done more than enough research for a thesis. The thesis needed to be written, the Ph.D. acquired, and a suitable position found. It felt quite strange. I was twenty-eight. Since the age of five, save for the war years (which seemed an unplanned, nightmarish interlude), I had been a student. Each year had followed the pattern of the previous in simple progression. Each year, I had known where I would be and what I would do in the next. The academic cycle had governed my life. I now came to an end of formal education, to a new and uncertain stage.

And now, even more startling, no academic positions appeared. The biology department received letters from departments in other schools around the country that were seeking faculty, but there was simply no demand for biophysicists, only for the more conventional zoologists or botanists or microbiologists. The possibility of unemployment had never occurred to me. To have done so well academically and then find there was no interest in my services was jarring. My problem was that I was a molecular biologist before there was such a discipline.

Another factor was that Joan and I were expecting our first child. In consequence of all of this, it was decided that I would stay on at MIT for another year as a (minimally) paid research associate to continue work on the microscope and ultraviolet spectroscopy projects and await a better season of job opportunities. During this year, Jesse and I wrote up the low temperature spectroscopy work and, after some editorial difficulty (what has this to do with biology?), saw it published in the *Journal of Biological Chemistry.*

All of these research projects may seem opportunistic, lacking a master plan. We were not at that time able to attack directly the key biological questions surrounding the nucleic acids. We were searching for a foothold, for ways to further exploit the scant clues we had, such as their specific ultraviolet absorption and its relation to the significant effects of ultraviolet irradiation.

The approaches we employed were novel. Each of the researches was a completely original foray into a previously unstudied area. Such studies are risky. They may produce important results or results of no value at all. But if well done they are definitive. Indeed, some twenty years later, a distinguished physical chemist told me that he had recently become interested in the absorption spectra of complex molecules at low temperatures and had been astonished to find that the best previous work had been done in the MIT biology department in the 1940s.

Of the three projects that comprised my graduate thesis, the reflecting ultraviolet microscope was later completed. It provided useful if not pathbreaking results. The low temperature spectroscopic studies of nucleic acids led no further. The study of the effects of ultraviolet irradiation on nucleic acid did open—directly and serendipitously—into much more fruitful paths, as will be seen.

By the following winter, at least two reasonable academic openings had appeared. One was in the biology department at Washington University in St. Louis—Frank Schmitt's former department. The other was a biophysics position in the physics department of Iowa State College at Ames.

I visited both places. Washington University had clearly changed little since the 1930s or even the 1920s, with a good but conventional biology department and little modern equipment or technical facilities. It did not seem well suited to my kind of biophysically oriented, instrument-dependent research. In contrast, several of the faculty of the physics department at Ames, including the chairman, Gerry Fox, had been at the Radiation Laboratory during the war. The theorists Julian Knipp and Frank Carlson and the solid-state physicist Gordon Danielson were also Radiation Laboratory alumni. The physics and chemistry departments at Iowa State had been involved with the Manhattan Project during the war, and now the AEC had established a permanent laboratory at Ames. A variety of excellent instruments was available, as were a fine machine shop and an understanding attitude with at least moderate resources.

In addition, there was a small but distinguished tradition of biophysics at Iowa State. Fred Uber, who had done the best studies of ultraviolet-induced mutation, had been in the physics department before the war but had moved to Missouri. And Iowa State was willing to recognize my Radiation Laboratory experience as a relevant part of my resume in deciding a salary level.

I was on my way to Ames.

Because few biologists came to biology with such background, I have been able throughout my career to bring a somewhat different perspective to bear on the problems of biology—an approach more quantitative, more analytically rigorous, more unrelentingly reductionist, perhaps more imaginative as to the ever-expanding potentials of ever-newer techniques. I became a laboratory biologist, but one always aware of a perspective external to the field.

Science, the Open Door

7

Iowa State—
The Young Professor

The class was over. I had concluded a discussion, with various examples, about electricity—about electromotive force and voltage, current, electrical resistance, and power. A young student came up and said, rather shyly: "I think I understand what you were saying, but there is one point I don't understand. Why do you need two wires to an electric light? Doesn't the electricity just flow in and get used up?"

Obviously, Iowa State was not MIT. I realized my discussion needed refinement.

Although my formal education had ended, Iowa State was providing other invaluable, on-the-job learning experiences. I learned how to teach in small classes and large, undergraduate and graduate; how to adapt to the mores of a large, public, state-supported university; how to establish and fund one's own research laboratory and program; how to advise graduate students, each with a unique background and personality; how to live in a small Iowa community.

Iowa State College (now University) of Agriculture and Mechanic Arts was one of the land-grant colleges established under the Morrill Act of 1862 to provide "practical" education. The more traditional liberal arts curricula and the professional schools of law and medicine were at the University of Iowa in Iowa City. Agriculture and engineering were the prominent programs at Ames, supported by a division of industrial science (mathematics, physics, chemistry, and biology) and accompanied by veterinary medicine and home economics. (Female students comprised about 25 percent of the undergraduates, mostly in the home economics program.) There were also supporting programs in the humanities, social sciences, and arts. Graduate programs were well established in agriculture, engineering, and the basic sciences.

The rather handsome campus had been designed for some six thousand students. Now, with the flood of returning veterans, enrollment had exceeded twelve thousand and was, in 1949, still at eleven thousand. Trailers for offices and classrooms dotted the campus and laboratories, and lecture rooms were used evenings, Saturdays, and even Sundays to provide all of the needed classes.

The returned veterans were older, more mature, more serious. They knew why they were at college.

Iowa State was in the top rank of the second tier of research universities in the U.S., except in agriculture, for which it was one of the very best. Its faculty included several members of the National Academy of Sciences and had high standards. Except for the agriculture department, which had considerable federal and state research support, the campus was limited by the available resources. The new Atomic Energy Laboratory at Ames was permitting a considerable expansion of the faculty and research activity in selected areas of physics and chemistry, but a broader lack of funds severely restricted the campus's ability to provide basic research equipment or to "ante up" matching funds for grants and imposed relatively large teaching duties. In consequence, the campus sought to recruit able young Ph.D.'s, nurture and exploit their bright ideas and enthusiasm "on their way up," and then reluctantly let them go, knowing it could not in general match the opportunities available at the leading research institutions.

Physics departments have a large "service" load, providing instruction in basic physics to students in many disciplines. All of the faculty, senior and junior, in the physics department at Iowa State participated in the teaching of elementary physics. It is a nice tradition, emulated in many physics departments, and one which I think would be salutary in other sciences such as biology or chemistry. This practice keeps all of the faculty in touch with the basics of their science and with the always changing flow of incoming students.

At Iowa State four different elementary physics courses were taught. The largest, of over one thousand students, was for engineers. Other, distinct courses were offered for science majors, for veterinary medicine students, and for home economics students. The physics faculty at that period was almost bipartite, split between the older professors, trained in classical physics before 1930, and the younger members, able and eager to apply quantum mechanical ideas to a variety of physical science questions. The former, now largely excluded from modern research, assumed primary responsibility for organizing the large service courses.

In addition to launching a research program with graduate students, I taught a graduate course in biophysics and two sections of one or another of the basic physics courses throughout the academic year. Although I had taken more than three years of physics courses at MIT, I quickly realized the truism that "you really learn a subject when you teach it." In order to teach a subject to others, one must thoroughly understand it oneself and must be able to adapt one's presentation to the initial level of comprehension of the students. Their questions can force unexpected insights. I acquired a much deeper grasp of basic physics through that experience.

Teaching sections of physics for the engineering students provided a special opportunity to appreciate the logistics of the provision of mass education. With nearly a thousand students and no lecture hall that seated more than 350, the demonstration lectures had to be repeated three times. I sat in on one of these to know what had been presented. With twenty-five students in each recitation session, there were forty such sessions in which homework assignments were returned and problems worked out, questions answered, and, if time permitted, further illustrations of physics presented. The instructors of all the sections met weekly to ensure coordination.

Two mid-term examinations and a final examination were given. In the days before machine grading, one instructor—sometimes me—graded one problem on a thousand examinations. There was inevitably some rivalry among instructors as to how well their sections would do on the examinations, but of course the statistical variations of student quality overwhelmed pedagogical ability. At the end of the quarter, all of the scores from examinations and homework assignments were summed and letter grades assigned. Since the cut-off points were arbitrary, a score of 891 might be an A and 890 a B.

Such mass education is economical, but it has many flaws.

This course provided my first contact with one of the corrupting influences in American higher education, big league athletics. One day I received a call from the football coach. A particular student in my physics section, who was doing poorly, was very important to the football team. Could I see my way to a more lenient grade so that he would be eligible to play next year? I thought quickly and responded that such actions were not my policy, but that I could recommend some physics graduate students who would be pleased to make some money by tutoring this student. This proved acceptable and the student's work actually improved significantly. I learned subsequently that such calls were not uncommon.

The course in biophysics, four lectures per week for thirty weeks, was a challenge. There was no text available. I attempted to analyze and present the physical principles underlying biophysical instrumentation (e.g., optical microscopes, the electron microscope, spectrophotometry, light scattering, X-ray diffraction, measurement of radioactivity, and so on), to discuss the biological effects of radiation (electromagnetic and ionizing), and to consider the physical properties of living tissue (primarily electrical). I had much to learn and I found my training in mathematics was often the essential key. The first year I averaged ten hours of preparation for each hour of lecture.

As might be expected, the undergraduates at Iowa State came mostly from the towns and farms of Iowa. In contrast to a highly selective private school like MIT, Iowa State admitted nearly every high school graduate in Iowa who applied. But then many could not achieve to the college's standards, and the freshman attrition rate was high.

Graduate students were drawn more broadly, although primarily from the Midwest. The college was obliged to accept a rather high proportion of graduates from the state's many four-year colleges, resulting again in a high first-year attrition (about one third of the incoming graduate students in physics dropped out). Only a few of the physics graduate students and some biochemistry majors (from the chemistry department) were interested in biophysics. Biology at Iowa State was unfortunately fragmented pragmatically into several divisions: zoology, botany, microbiology, and genetics. Other subdivisions of economic importance (e.g., entomology) were located in agriculture. For most of the biology faculty, biochemistry and biophysics were arcane subjects, and their undergraduate curricula required neither calculus nor physics.

I felt strongly that the lack of such requirements was a grievous deficiency that would handicap the biology students throughout their future careers. But, despite repeated efforts over the years, I was unable to effect a change.

Agriculture was the primary locus of biological research funding on the campus. I was fully confident that research on nucleic acids, such as I was undertaking, would be key to the long-range future of agriculture, and I attempted on several occasions to persuade the dean of agriculture of this view. He listened attentively but was never persuaded. In truth, agricultural research of the conventional kind was doing very well in that period and agricultural yields were increasing yearly. Twenty-five years before recombinant DNA, my enthusiasm, albeit justified, was, pragmatically, premature.

Iowa in 1949 was primarily a rural state with no large city and was composed mostly of farms and small, relatively isolated communities. It was little changed socially from 1910 or 1920. Alcohol could be purchased only at state liquor stores in the county seat. The sale of margarine was illegal. Many major issues of the time (the Rosenberg trial, the Hiss affair) penetrated little into our consciousness.

Life in Ames had an even, pleasant tenor. It was a small community, a university town, with seven to eight thousand permanent residents and eleven thousand students. Ames was small-town America, lacking the extremes and diversity of a city—there was no great wealth and no real poverty. Crime was almost nonexistent and doors were seldom locked. When I arrived, there were no stoplights, no large stores, few (and mediocre) restaurants. There were no "minority problems." There was a genuine small-town kindliness and neighborliness. As might be expected in a university town, the public school system was excellent. It was an ideal place to raise children.

There was almost no rental housing in Ames. All of the permanent residents lived in their own houses. Most of the students were housed on campus, the women's dormitories on the east side, the men's on the west. To accommodate the influx of married students, a sea of Quonset huts named Pammel Court had been erected—each hut was made into a small two-bedroom house. We lived there for our first two months. Having just completed graduate study, I could relate to the concerns of the surrounding students, even though I now had a different status.

We then located a frame house, about thirty years old, four blocks from the college on a dead-end street. The house backed onto a college experimental farm and was two blocks from elementary school. My father loaned me two thousand dollars for the down payment, and a 2.5 percent FHA mortgage of sixty-five hundred dollars covered the rest. My annual salary was fifty-six hundred dollars, out of which four hundred dollars was deducted for the retirement plan. The remainder was enough to get by on in Ames in 1949.

The cultural life of a small Iowa community can be quite minimal, but in Ames the presence of the college provided a wealth of activities. In addition to the plays and concerts that the campus itself generated, there was a visiting musical artist series, including symphony orchestras, and there were guest lecture and foreign film series. And of course there were the athletic contests, especially football and basketball, and a great student-organized fall harvest/football homecoming festival, VEISHEA (Veterinary medicine, Engineering, Industrial Science, Home Economics, Agriculture), with a parade of fanciful floats.

Television was just then making its first incursions into this semirural idyll. We acquired our first TV set, a small black-and-white model. We had resisted this "invasion" of the outer world for years, but we found that the children were disappearing to the neighbor's houses to watch. Also, references to TV programs of which our children were ignorant were made in school classrooms. TV had already become an integral part of the culture. To my great surprise, I was, initially, mesmerized. To be able to watch movies in one's own home had been an undreamed of luxury, available only to movie magnates and presidents. Now we had this privilege! Of course, the fascination wore off, and after a time I could restrict TV to a reasonable share of my existence. Indeed, familiarity breeds contempt. As I became accustomed to this technological marvel, I became increasingly selective and guarding of my time. Today with cable we have thirty-six channels—and often nothing I want to watch.

Iowa State provided my first introduction to campus politics—the intricate network of relationships and personalities that parallels, and sometimes displaces, merit in the allocation of resources and rewards. Iowa State had conventional hierarchical lines of authority. Mine ran through the department chairman to, variously, the dean of science or the director of the Ames Laboratory of the AEC (which provided much of the research funds available to physics) and (of limited access) through the dean to the president. The Ames Laboratory had no mandate for biophysical research. Thus, while I was able to "bootleg" shop time and instrument repair from their well-maintained facilities, I received no direct support from that source.

The dean had no budgetary item for research equipment. Iowa operated on a biennial budget. If, as the end of the biennium approached, unexpended funds remained in the college budget, the dean received a sum that he could use to meet the numerous requests as he believed best. He was clearly influenced in this regard by his personal relationships with the department chairmen. It was also widely recognized that he hoped to be the next college president and his allocations reflected judicious efforts to strengthen his political base, at least on the campus. (He did not succeed.)

It soon became very clear that, if a faculty member had a personality clash with his department chairman—who was appointed by and was often a friend of the dean—his situation on the campus was hopeless. Regardless of academic merit, he could be denied resources, limited in laboratory space, assigned excessive teaching responsibilities, passed over for promotion, denied increase of salary, and so on. And this happened, painfully, to some. In one especially grievous instance, I attempted to intervene at the dean's level and was somewhat curtly rebuffed.

Fortunately, my relations with my department chairman were excellent. And indeed, while some of the physics faculty probably were perplexed as to the significance of biophysics, in general my relations with the other faculty were warm, and close friendships developed between families. Of course, it may have helped that that department was rather well supported and I was not a competitor for their funds.

As a public institution, Iowa State was subject to a variety of regulations and policies more appropriate to a department of highways or a water commission than to an institution of higher learning. Purchasing, accounting, and time-reporting procedures were rigid and occasionally stultifying. Indeed, there was at times a sense that we were public servants, "hired hands," available at the public's call to perform various functions to their satisfaction.

A tragedy that occurred in my first year at Iowa State has left a sharp, poignant memory. Frank Carlson, a theoretical physicist and a student of Robert Oppenheimer's, had been with me at the Radiation Laboratory and was now on the faculty at Ames. A thoughtful, sensitive, private man, we quickly became good friends. One morning, I saw him standing in the hallway outside the physics department office with a dazed, distressed look as swarms of students swirled by him, to and from their classes. He was motionless and I thought to push through to him to ask if he were all right, but I had to hasten to my class and thought I'd speak to him later. I tried to find him later that day, but he was not in his office. Next morning, we learned that Frank had hanged himself in his basement. He had been increasingly depressed for some time; psychiatric help had been of no avail.

In our small world, this was a searing shock—and I felt, and still feel, a deep remorse that I had not followed my instinct to reach him. Since, I have always responded at once to a look of distress. There may not be a second chance.

It soon became evident that I would have to find an outside source of funds for my research. The department provided some initial "start-up" equipment, but money for supplies, chemicals, a laboratory assistant, and so on would have to come externally. I intended to pursue my research into the nature of the effects of ultraviolet radiation on the nucleic acids. I knew the Rockefeller Foundation had been interested in John Loofbourow's ideas at MIT so I applied to them. With, I suspect, Loofbourow's recommendation, I received a grant of five thousand dollars per year for two years. I could get underway!

Toward the end of my first year at Iowa State, I saw a notice that a Gordon Conference—an intense, week-long gathering of experts—was to be held that summer on the subject of nucleic acids and proteins at a small preparatory school in New England. This seemed a fine opportunity to catch up on the most recent findings in the field. Jesse Scott and I both applied to attend and were accepted. This meeting would also provide my first opportunity since leaving MIT to visit with John Loofbourow in Cambridge on my way to the meeting.

A few weeks before the meeting, I received a shocking telegram. John Loofbourow had died suddenly of a massive heart attack. I felt a deep and continuing loss. John had been my mentor, guide, scientific role model, friend. We would never again share happy hours of discussion and scientific inspiration. I am sure this early and tragic loss of my mentor significantly influenced my subsequent career. I know there have been numerous times I would have liked his advice, his calm wisdom. And of course, in the network of science his backing could have provided forms of access that I subsequently lacked.

John's wife, Dorothea, was grief-stricken and, one week later, she too died of a heart attack. A stunning blow.

The Gordon Conference went on as planned. It was a remarkable event. There were some ninety-five participants who included almost everyone who was doing physico-chemical or structural or metabolic studies on the nucleic acid or protein molecules—Paul Doty, John Kendrew, and many others. Since 1950, this field has expanded again and again. For simple reasons of size, the researchers of nucleic acids and proteins had to divide into separate conferences by the early 1960s, and the nucleic acid sessions soon began to alternate, as between structural studies and metabolic and functional studies. Today only a small sector of each field can be discussed at any one conference.

8

Iowa State—
Research and Discovery

In 1949, much of the physics and chemistry of life was still utterly mysterious. Ill-defined, large molecules called proteins apparently catalyzed most essential chemical reactions and provided structural and contractile components. The origin—the manner of synthesis—of these proteins was completely obscure.

Several lines of evidence implicated the nucleic acids in genetic processes and suggested that they might indeed convey genetic information. If so, they had to play a primary role in protein synthesis. But knowledge of the structures of the nucleic acids was even more tenuous than that of proteins. We knew that there were two broad classes: deoxyribonucleic acid (DNA) and ribonucleic acid (RNA). Both were chains of monomers, nucleotides. They differed chemically in the sugar portion of each component nucleotide, deoxyribose in DNA, ribose in RNA. In addition to common nucleotides, each contained a distinctive nucleotide, thymidylic acid in DNA, uridylic acid in RNA. All cells contained both DNA and RNA. Some viruses contained DNA, some RNA.

In contrast to the proteins, some of which had even been crystallized, no one had ever obtained a pure, single DNA or RNA species or indeed, as far as was known, a complete "molecule." Only complex mixtures, mostly fragments, were available. And, if a pure, single species had been available, there were no means to analyze its composition at the nucleotide level or to characterize its functional role. Nothing was known about the synthesis of nucleic acids or of their metabolic turnover, if any. Yet these now appeared to be the carriers of inheritance.

In science, the choice of a specific field of research and a means of approach is critical. When much is unknown, it is very difficult to forecast which approaches or paths will be fruitful and which will be deadends. My choice—to explore the domain of the nucleic acids and to seek to elucidate as much as I could of their structures and, in time, their functions—was most fortunate. The nucleic acids, which had occupied a minor and obscure niche in biochemistry, were about to move to the center of the biological stage for the next several decades. So little was known about the nucleic acids that almost any new insight would be valuable. And, in 1949, there were few competitors in this biochemical backwater.

With my background at the Radiation Laboratory and in spectrophotometry, and located in a physics department with access to (then) modern spectrophotometric equipment (for infrared and Raman spectroscopy), and provided with graduate students who had backgrounds in physical science and were familiar with physical instrumentation, it was natural for me to seek to explore the structure of nucleic acids by means of their interaction, in various modes, with electromagnetic radiation—to study the absorption, re-emission, and scattering of infrared, visible, and ultraviolet radiation by nucleic acids and to look for photochemical changes induced by such interaction. Studies of this nature had been very powerful means of elucidating the structures of simpler organic molecules and it seemed reasonable to believe they would provide a valuable approach to the study of nucleic acid structures.

We knew that DNA molecules were long, polymeric chains of deoxyribonucleotides. Four different subunits were known—adenylic, guanylic, thymidylic, and cytidylic. Each was composed of a complex ring structure, either a purine (adenine or guanine) or a pyrimidine (thymine or cytosine), joined to a sugar (always deoxyribose), joined in turn to a phosphate group. The subunits were linked together through the phosphate groups between the sugars. The purines and pyrimidines were strong and distinctive absorbers of ultraviolet light.

For the research I wanted to undertake at first, it seemed preferable to study the true nucleic acid subunits, the deoxynucleotides, rather than the isolated purines and pyrimidines. However, in order to do this, I needed a source of deoxyribonucleotides. They were not commercially available then and for good reason—the best available methods for their isolation provided yields of less than 1 percent from a DNA source. Nor was an undegraded DNA of good quality then commercially available.

One of the major heuristic differences between the biology of the

1950s and the biology of the 1990s is the presence today of numerous small companies that supply, for a price, a wide variety of the chemicals, enzymes, even viruses and cells employed in biological research. This development, largely concentrated in the United States, has greatly accelerated the pace of biological investigation. In the 1950s, one had to prepare one's own DNA, as well as the various enzymes needed to digest or modify it, starting usually from fresh animal tissue obtained from a slaughterhouse. Each preparation was not only time-consuming but often involved an interim research project to define the optimum conditions for each step or to apply new techniques that hopefully would provide more acceptable yields.

By preparing our own DNA, purifying our own enzymes, and introducing new fractionation methods, we were able for the first time to digest DNA completely to its component mononucleotides and to prepare these quantitatively. In addition to the four principal mononucleotides we expected (adenylic, thymidylic, guanylic, and cytidylic), we found a minor fifth mononucleotide in our digest that proved to contain the then recently discovered pyrimidine, 5-methylcytosine (a modified form of cytosine).

The mononucleotides were then used for the planned ultraviolet irradiation studies, as well as for other research such as pioneering Raman (the study of the spectra of re-emitted radiation) and infrared absorption spectroscopy. This information could establish the tautomeric state (i.e., which of several possible alternative atomic groupings was predominant) of the various substituent groups on the purines and pyrimidines. The ultraviolet irradiation studies demonstrated that, on exposure to ultraviolet, the cytidylic acid underwent a reversible loss of its ultraviolet absorption similar to that we had previously described for uracil and uridylic acid but more rapidly revertant. The other deoxynucleotides were more resistant to ultraviolet irradiation.

The infrared absorption studies were performed with solutions of the nucleotides. These experiments were complicated by the fact that water itself has strong absorption bands in regions of the infrared spectrum. To circumvent this difficulty, we also studied the infrared absorption of the nucleotides dissolved in heavy water (D_2O), as its absorption bands are shifted relative to those of H_2O. These studies permitted the assignment of many absorption bands to specific atomic groups in the deoxynucleotide molecules and established that, under biological conditions, they are in what are known as the keto and amino configurations, as opposed to the alternative enol and imino configurations.

These chemical distinctions are very important in determining the kinds of secondary bonds that can then be formed between the nucleotides in higher level structures. The determination of these structures was confirmed by the results of the Raman spectroscopy. In those days, before lasers, the latter experiments were very difficult to perform because of the low light levels available.

It was known that some viruses, such as the tobacco mosaic virus, did not contain DNA but had the other form of nucleic acid, ribonucleic acid, RNA, as their (probable) genetic material. I therefore wanted to study RNA. RNA is abundant in cells, but at that time it could only be isolated as a complex mixture. To obtain a pure, individual RNA, I therefore undertook to grow and isolate the tobacco mosaic virus. After all, Iowa State was an "Ag school" and as such had well-equipped greenhouses. The authorities were at first unenthused about my growing stocks of infected tobacco plants in their greenhouses (the virus can also infect such plants as tomato), but reluctantly they consented.

The tobacco mosaic virus had been isolated and purified to a quasi-crystalline state by Wendell Stanley in the 1930s. The dimensions of the virus particle had been determined in the electron microscope. It had a particle mass of approximately forty million daltons. By chemical analysis, it was known to contain about 5 percent RNA. However, it was quite unknown whether there was one RNA molecule (of mass two million daltons) in each particle or a set of several smaller molecules, which might or might not be identical. We sought to resolve this question by performing light-scattering studies on the isolated viral RNA. We built and calibrated our own light-scattering apparatus. By measurement of the light scattered at various angles from a solution of macromolecules, one can determine the number of scattering particles and therefore, knowing the concentration of the solution, their molecular mass. One can also, assuming a structural model, determine their spatial dimensions.

We introduced a new concept. By measuring the scattering at different wavelengths in the visible and ultraviolet, we were able to differentiate between various possible structural models to determine, unambiguously, the spatial dimensions of the RNA. Our results demonstrated that, when carefully prepared, the RNA was a single molecule of mass two million daltons.

We were, step by step, establishing solid facts about the nucleic acids and their components: absolute absorption coefficients, confirmed structures, molecular masses and dimensions. We needed, however, a more direct link to function.

In the 1940s, microbiological assay and chromatographic methods are developed that permit the quantitative determination of the amino acid composition of proteins. Arthur Martin, Richard Synge, William Stein, and Stanford Moore play leading roles in this development. Building upon these methods, Fred Sanger develops techniques for the determination of amino acid sequence within peptides and over the period 1951 to 1955 determines the first complete amino acid sequence of a protein, insulin. With this advance, proteins, with all their critical biological functions, can be regarded as complex but precisely definable organic molecules.

In 1950, Linus Pauling and Robert Corey propose the alpha-helix and the beta-pleated sheet models for the spatial arrangement of amino acids in proteins. These structures, since verified as components of numerous proteins, provided the first valid concepts of the spatial organization of proteins.

In 1952, Alfred Hershey and Martha Chase demonstrate that DNA is (most likely) the genetic component of the bacteriophage T2, i.e., that DNA (without protein) can be the exclusive carrier of genetic information.

9

Max Delbrück—
A Caltech Interlude

For an audience of generally reserved physicists, the applause was extraordinary. Max Delbrück had just concluded his third lecture on his path-breaking research with bacteriophage. These studies, over a fifteen year period, had brilliantly transformed a microbiological oddity into a principal means for the advancement of molecular biology, genetics, and biophysics.

The lectures were stirring. The research, the clarity of its conception, the elegance of its execution and interpretation, and the lucidity and sophistication of its presentation caught everyone's imagination and enthusiasm. Biology, a messy and descriptive field to most physicists, could be approached in a quantitative, analytical manner often using mathematical insight to produce clear, unambiguous results.

Frank Spedding, the director of the Ames AEC Laboratory, had made funds available to invite distinguished visiting lecturers. My nomination of Delbrück had been approved, and to my great delight he had accepted the invitation to stop over on his way to some East Coast meetings. We had never met. Delbrück was tall, thin, and crew cut with an energy that belied his forty-five years. In his three days, he met with most of the more prominent research faculty and, I believe, was pleasantly surprised by the activity he found on his first visit to a midwestern "Ag school."

I had become increasingly convinced that my biophysical and biochemical studies on nucleic acids would be much more meaningful if they could be related to a functional biological process. The long-range goal was to understand what nucleic acids did, how they did it, and how their activity was controlled. The bacteriophage system as developed by Delbrück seemed ideal for my purpose. After his visit in the fall of 1951, we remained in contact. When informed of my interest, he proposed that I should come to his laboratory at Caltech to learn the current state of bacteriophage biology and associated techniques.

In my third year at Ames, I was not yet eligible for sabbatical leave. However, I could take six months unpaid leave. I sought fellowship support for the six months. After my interview, I was informed by the fellowship board that they would award me a fellowship not for six months, which they thought would provide inadequate training, but for one year! My dilemma was happily resolved when Max was able to offer me a six-month fellowship from Caltech itself. I was to start on 1 January 1953.

We left Ames before Christmas in the midst of a sleet storm, wended our way across the frozen Midwest to Texas and on to Arizona, and spent Christmas in the snow at the Grand Canyon. Finally, we crossed the Mojave and descended through Cajon Pass into southern California. Once again, I had the sudden shock of emerging from bitter winter into warm sunshine and palm trees and fragrant flowers. We were just in time for our first Rose Parade.

Caltech was a revelation. Here were "giants"—scientists of the highest eminence—in almost every field. Lee DuBridge, who had directed the Radiation Laboratory, had assumed the presidency of Caltech in 1946 and had revitalized the institute. He had brought George Beadle, the leading geneticist, to head biology, and along with him Norman Horowitz and Herschel Mitchell. Alfred Sturtevant and Ed Lewis continued Thomas Hunt Morgan's drosophila research. Frits Went and James Bonner led an outstanding program in plant physiology. And Max Delbrück, Renato Dulbecco, and Jean Weigle were preeminent in bacteriophage research.

In chemistry, Linus Pauling was preeminent in the field of molecular structure, aided by Bob Corey. Jerry Vinograd, Walter Schroeder, and Norman Davidson were other leading chemists in the Pauling orbit. Jack Roberts had recently arrived to launch his NMR (nuclear magnetic resonance) program. In physics there were Richard Feynman, Murray Gell-Mann, Carl Anderson, and Charles Lauritsen; in geology Charles Richter and Benno Gutenberg; in astronomy Jesse Greenstein, Horace Babcock, and Martin Schmidt; in engineering Clark Millikan, Hans Liepman, and so on.

In a large university, a "post-doc" in Delbrück's laboratory such as myself might never have even seen these great men. But Caltech is a small, surprisingly democratic institution, and I met all of them and had significant discussions with several. Caltech was clearly a world center for science, one of the leading two or three in the world. In biology, almost everyone of distinction visited and lectured periodically. The newest discoveries were quickly known and discussed in a ferment of ideas. Biology at MIT had been much less dynamic, and the contrast with the near-isolation of Ames was extreme.

Max Delbrück was a profound influence on many students and colleagues, largely through the sheer force of his personality. Bright and rigorously logical, he imposed a quantitative intellectual discipline on the field of bacterial virology, which had been largely qualitative and unfocused. As the founder of a revived bacteriophage research program, Delbrück served as a combination pater familias and Herr Professor for the informal college of phage researchers developing about the country. He had insisted that they exchange cultures and limit the number of different bacteriophages studied so that different laboratories could obtain comparable results. Any new findings were quickly communicated to Delbrück for his comment and evaluation and through him to the community. He had a remarkable status and for a while produced an extraordinary cohesion within the phage fraternity. Of course, in time the field outgrew its founder.

In scientific debate, he could be mercilessly caustic but never mean-spirited. Delbrück had a weekly phage seminar at which ongoing research was discussed, or a visitor lectured. Max was known for his blunt and open criticism, his insistence on clear concepts and logical presentation, and his persistence in exposing ambiguities and uncertainties in a speaker's chain of reasoning. Many a speaker in his seminar or in the general biology seminar was sharply deflated; however, the net effect was to set and enforce a high standard of presentation.

His approach with students was generally "sink or swim"; it was salutary for the more resourceful, devastating for others. At the same time, Max was fond of playing the father-professor role, throwing large and often elaborate parties for the phage group at his house near campus, and taking caravans out for camping trips on the desert. The desert was quite unpopulated in those days and even a large group could be quite isolated in its own valley. These excursions featured long hikes and climbs and explorations of caves. At night, the brilliance of the stars in the clear air, absent city lights, was startling to one raised in the always slightly hazy Midwest and East. Lying out in the open and watching the heavens revolve enabled me to comprehend how significant and mysterious the nightly stars must have seemed to primitive man.

Max had several "Maxims." For delivering a lecture to an audience of uncertain background knowledge, his advice was to "assume they are totally ignorant and infinitely intelligent." When skeptical of a new result, he would say, "I don't believe a word of it."

Max was convinced that important principles should be explicable in a simple and straightforward manner. If a convoluted and arcane argument was used, he was dubious. His comment "what a swindle" indicated his belief that some big step had been overlooked or some error had been made in the complex exposition. He was frequently bemused by the progressivity of evolution—that each evolutionary advance, presumably achieved by selection to cope with a particular environmental circumstance, seemed again and again to bring additional new and unexpected potentials. "Nature provided more than was needed." The T4 bacteriophage he had studied proved to have over 150 genes, many of which would be of significant utility only in quite exceptional circumstances. The evolutionary steps that converted our anthropoid ancestors into *Homo sapiens* could hardly have been intended to solve problems in X-ray diffraction or quantum mechanics.

Delbrück's own research had mixed success. A physicist by training and inclination, he had little taste for the complexities and variations of chemistry. He sought out problems that could be approached quantitatively, analyzed abstractly, and preferably studied with simple equipment. This bent succeeded admirably in his early studies of bacteriophage, in the quantitative analysis of mutations and the discovery and analysis of bacteriophage genetics, and in his insistence that workers in bacterial virology should focus their efforts on a limited set of viruses.

But this approach was much less successful in his later studies of the biological effects of ultraviolet radiation on bacteriophage—in retrospect, the analysis was hopelessly complicated by the existence of multiple cellular repair mechanisms—and in his prolonged effort over twenty years to establish the mold *Phycomyces* as a model system for the study of sensory processes. This last was doomed by the inability to develop *Phycomyces* genetics. At the symposium in celebration of fifty years of biology at Caltech, Max, as chairman of a session, gently chided Dale Kaiser for having spent too many years researching the lambda bacteriophage, long after the more significant discoveries had been made. Dick Feynman, sitting next to me, leaned over to say that Max should have applied his advice to his own work on *Phycomyces*.

As a physicist, Max sought to find principles of great generality, as opposed to solutions involving specific chemical interactions. It was his initial hope that the phenomena of genetics—cloaked in obscurity in the 1930s—would prove to rely on principles of physics previously undiscovered. In this, of course, he was grievously disappointed.

Much later, one of the most pleasant activities of my tenure as chairman of biology at Caltech was to organize the celebration for Max after he received the Nobel Prize in 1968. Admired, even exalted, for two decades at the institute, Max was felt by all there to deserve the Nobel Prize for his bacteriophage work. But after he had been passed over for so many years, and biology had moved on, many feared it was not to be. This made for all the more delight when the word came.

Max took the ensuing hoopla with his usual diffidence. I well recall the news conference the next day, with Max attempting to display and explain his current work with *Phycomyces*, which was completely unrelated to the research for which the prize was awarded. Prominent among the newspeople was a TV reporter who did a nightly interview with a local personality. The evening before, she had interviewed the superintendent for sewers. (Science reporting was at a low ebb in those days.) Actually, while Max was very pleased at the belated recognition, he was a little embarrassed that it had come long after he had left the phage field. He could no longer cogently discuss the current state of the science; bacteriophage research had passed him by. Knowing this, he would deprecatorily call himself "the old windbag" at the inevitable required speeches.

The celebration party took the form of an elaborate musical skit with various members of the biology division commemorating and commenting in humorous doggerel on the several phases of Max's career. It was a loving "toast and roast" of a deservedly admired and venerated presence who had finally been appropriately honored.

At Caltech, I was primarily a student. I learned phage techniques by working through a lengthy set of prescribed laboratory exercises—I recall Max watching amusedly as I first learned to master sterile technique, that is, the art of holding all the tubes and cotton plugs while transferring liquid from one container to another. I also learned the current state of phage research in the group discussions. But I learned far more of biology and chemistry by attendance at the great variety of seminars and from conversations with visitors and Caltech faculty and researchers, almost osmotically absorbing the ambience. Those six months introduced me to science at a still higher level of intensity and synergistic interaction.

In 1953, James Watson and Francis Crick propose the double helix structure of DNA, based upon the X-ray diffraction studies of Maurice Wilkins and Rosalind Franklin. This structure made explicable in chemical terms many of the previously mysterious properties of genes and opened the way to molecular genetics.

10

A Sidelight on Watson and Crick
(A Corollary to Murphy's Law: Seekers/Finders >> 1)

The elucidation of the double helix structure of DNA by Watson and Crick has taken on mythic dimensions. The authors of the myths, non-scientists and (even) scientists like Watson himself, at the time ignorant of the history of DNA research, have presented a sometimes self-serving scenario akin to the primitive myths of creation in which the world—or the DNA structure—is derived from a formless void. Of course, it wasn't like that.

Science today is a cumulative enterprise rising step by step on a staircase of four centuries of sustained investigation and analysis. Each step has a background and a context. Indeed, the quick acceptance of the double helix structure derived from the fact that it made coherent and even plausible so many prior observations from genetics and DNA biochemistry.

As a participant in DNA research throughout this period, I had a close-up and somewhat bemused view of the ongoing research. As is the norm in science, several lines of research, including my own, were converging on the structure of DNA. The concept then of the double helix was, in itself, not so large a step—but it was a step that brought biology onto a new plateau with wide horizons and numerous paths to explore.

By the early 1950s, several laboratories were intensively engaged in the study of DNA. Arthur Kornberg's laboratory, then at Washington University and later at Stanford, sought to unravel the biochemistry of

DNA synthesis. Paul Doty's laboratory at Harvard applied powerful tools to the analysis of the peculiar physical chemistry of DNA. Alfred Mirsky at Rockefeller Institute had measured the DNA content of cell nuclei of varied species while Erwin Chargaff had shown that the purine and pyrimidine composition of these DNAs always displayed a curious regularity: the moles of adenine equaled those of thymine and the moles of guanine equaled those of cytosine. In London, Franklin and Wilkins sought to improve on the earlier X-ray diffraction studies of DNA by William T. Astbury.

I too, by a somewhat circuitous route, had become interested in aspects of DNA structure and sequence. In the course of my earlier studies, intended initially solely to provide the mononucleotides I needed for irradiation studies, I realized I had also opened an avenue into questions of nucleotide sequence and DNA structure. For in the course of the mononucleotide preparation, we at one stage reduced the DNA to a mixture of small polymers of an average length of four nucleotides. If we could fractionate this mixture we would be in a wholly unexplored realm, since until then no one had ever isolated, identified, or purified oligonucleotides (short strings of nucleotides) of any size or variety.

We found that we could indeed separate the digest into dinucleotides, trinucleotides, and larger fractions. And further, by careful choice of conditions, we could fractionate all of the dinucleotides according to their particular nucleotide composition. The ultraviolet absorption of each dinucleotide suggested its composition. This could be verified by degrading each to its mononucleotides and fractionating these.

This work proceeded well and brought us, for the first time, a small step into the world of DNA sequence—the order of nucleotides in DNA. By midsummer 1952, we knew that:

(1) The unusual nucleotide 5-methylcytidylic acid was to be found only in one isomeric form of one dinucleotide, which meant that it always preceded a guanylic in the DNA chain (this has proven to be general in the DNA of mammals and the presence of methylcytidylic acid is now believed to be involved in the control of genetic expression);

(2) All of the other two-nucleotide combinations could be found in reproducible but varying proportions.

This latter was quite significant. One possible explanation of the Chargaff regularities (moles adenine = moles thymine and moles guanine = moles cytosine) had been that perhaps adenylic was always followed by (or preceded by) thymidylic, and guanylic similarly by cytidylic. However, this clearly was not the case. The Chargaff regularities demanded another explanation.

I discussed this question with Fritz Schlenk, then professor of microbiology at Ames. The simplest explanation I could propose was that there must be two DNA chains, related in some way, so that wherever there was an adenylic in one chain, there was thymidylic in the other, and likewise for guanylic and cytidylic plus methylcytidylic. This hypothesis had an obvious consequence, but unfortunately I was not then in a position to test it. If it were correct, then for every dinucleotide sequence, say CA, there should be an equimolar amount of the complementary sequence GT from the other chain (or TG if the two chains should have opposite orientation).

But our dinucleotides amounted to only one-sixth of the partial digest, and we could not then sequence the larger oligonucleotides. Nor could the dinucleotides present be considered a random sample as we were ignorant of possible preferences in the enzymatic digestion. Indeed, biochemical verification of this concept had to await the development of the "nearest neighbor" technique by Arthur Kornberg in his later studies of DNA synthesis.

At that time, I did not think in terms of helical structures and while I thought of the DNA as genes, I could not suggest any good reason for the presence of two such chains other than vague ideas about possibly coding for two proteins with some oddly defined relationship. My thoughts were oriented toward gene structure and expression, rather than toward gene reproduction and mutation, and so I missed the true significance of the complementary chains (so obvious when they are paired). This oversight was the result of an education that emphasized biochemistry and biophysics rather than inheritance and variation.

Shortly after I began my fellowship at Caltech in January 1953, I presented our dinucleotide data in a biochemistry seminar. I was somewhat surprised at the intense interest shown. I mentioned the two-chain notion, but, remarkably, the focus of attention was on the quantitative relationships among the dinucleotides, even though these accounted for but a fraction of the digest. I soon appreciated that the reason for this interest lay in prior discussions at Caltech about possible coding schemes to relate DNA sequence to protein sequence. Ours was the first quantitative data about any aspect of nucleotide sequence.

In February 1953, Linus Pauling announced a seminar at which he would present the structure of DNA as deduced from X-ray diffraction data. As Pauling had earlier discovered the basic structures of proteins from X-ray diffraction data, his seminar was eagerly awaited. He put forward a triple helical structure for DNA with the phosphate groups on the inside and the rings of the nucleotides facing out. This structure was met with much skepticism. Many of the chemists present felt that the charged phosphate groups could not possibly be located in such close juxtaposition in the interior of the molecule.

Max Delbrück and I went to see Pauling to suggest possible modifications. Max was concerned that the chains of the helix should be paranemic (i.e., combable) rather than plectonemic (i.e., intertwined) so that they could be separated without the complication of unwinding—a very real problem for which the cell has had to devise special mechanisms. I wanted to propose that Pauling examine a two-strand helix to incorporate the Chargaff regularities and my notion of two complementary strands.

Pauling, however, could not be persuaded. He felt that his model was the only solution to the X-ray data and the known density of DNA. Unfortunately, as it later developed, he was using the old X-ray data and density data of Astbury, which had been obtained with a poor preparation of DNA that had been significantly denatured. Pauling had no access to the newer X-ray diffraction data of Franklin and Wilkins, which undoubtedly would have quickly set him on the right track.

The following month, Delbrück received a letter from Watson setting forth the Watson-Crick model. This immediately resolved many of the concerns raised about the Pauling model, which was abruptly discarded. The Watson-Crick model was then presented with immediate acceptance at the Cold Spring Harbor Conference in 1953. The convincing evidence for the biological, as opposed to the chemical, reality of the Watson-Crick model came from the Meselson-Stahl experiments at Caltech in 1957 demonstrating the separation and conservation of each of the DNA strands on replication.

The only substantial, if temporary, challenge to the Watson-Crick model came with my later discovery of the single-stranded DNA of the bacteriophage, φX. If Watson and Crick were right, how could this DNA reproduce or even code? The conundrum was subsequently resolved with the demonstration that, within the infected cell, the single-stranded viral DNA was quickly converted to a double-stranded replicative form, which replicated as such and which coded for proteins.

Later in the infective cycle, viral-coded proteins produced the single strands of DNA for progeny viral particles, using one strand of the double-stranded replicative form DNA as a template.

The Watson-Crick structure, by providing a firm biochemical basis for genetic phenomena, led biology into a new era, but it did not emerge unprecedented from a void. Mine was but one approach. Other lines of research in the laboratories of Kornberg, Doty, and Franklin and Wilkins were converging on the double helix. The double helix was out there waiting to be revealed. By 1950, after a century of biochemistry and fifty years of modern genetics, it was but a short step further in the unknown.

Science is, in fact, a collective enterprise. To outsiders, however, Watson and Crick by virtue of their great discovery and their strong personalities have made it seem a more personal adventure.

Jim Watson is the stuff of which *People* magazine is made. Brilliant, arrogant, verbally crude, with a skewed, off-center personality, fond of publicity, he sees the world in black and white with little gray in between. In his career, he has consistently demonstrated excellent scientific judgment and has been a superb director of the Cold Spring Harbor Laboratory. Crick is an interesting complement—highly articulate, sometimes glib, skilled in debate, sophisticated, and seemingly self-confident, even arrogant, yet harboring a persistent personal reserve.

Watson's book *The Double Helix* was published in 1968. Seemingly an exercise in candor, it portrays him as a total opportunist, devoid of any sense of scientific community, heedless in his passion to "beat Pauling" in the race to the "golden prize." Of course, Pauling never knew he was in a race. As a testament, it conveys an impression of science as just another cutthroat competition, akin to Wall Street or even the political arena. Shortly after the book appeared, I was awarded the California Scientist of the Year prize by the California Museum of Science and Industry for our research in collaboration with Arthur Kornberg at Stanford. At the inevitable news conference, reporters—with Watson's account fresh in their minds—could not comprehend the genuine, and really quite routine, collaboration between our two laboratories in the common effort to advance scientific knowledge.

11

Iowa State—At Full Speed

It was a rare privilege. Robert Oppenheimer and I were having lunch. He had come to Ames to deliver the first Frank Carlson Memorial Lecture, to honor his former student. We spoke of many things: the Institute for Advanced Study, which he now headed; the state of theoretical physics; the international situation; his long-standing interest in Hindu philosophy. He was curious as to the state of biophysics. I avoided mention of his recent "trial" at the AEC and his loss of security clearance—in my view a gross miscarriage of justice. He seemed to bear no rancor and his general mood was philosophical, detached, with an eye to the long view of events. I was captivated by his use of language— eloquent yet precise, almost poetically phrased.

I did not fully comprehend his subsequent lecture, which concerned calculation of the self-energy of the electron and discussed more recent refinements of work that Frank Carlson had initiated as Oppenheimer's student. But the luncheon discussion was a glimpse of true and wide-ranging brilliance.

After Caltech, my return to Ames was bittersweet—a homecoming, yet a letdown. Almost immediately, I went East to the Cold Spring Harbor Conference, at which the double helix structure for DNA was formally presented by Watson and Crick. After Caltech and Cold Spring Harbor, Ames seemed quiet—and, now, isolated.

As quickly as possible, I established my phage research. The bacterial viruses provided enticing access to a wealth of important biological problems. With them, genetic structure and function, gene replication and mutation were all open to investigation with the several techniques we had been developing. These techniques were not perfect; they were limited and often tedious, but they would take us a long way. The full fruition of this was to come later with the research into the small bacteriophage φX174, but for the present I undertook to analyze the nucleic acids of the well-known T-series of phages.

There was a quiet moment of triumph when I observed my first phage plaques in Ames. My facilities were not comparable, but I had transplanted the phage technology from Pasadena to Ames.

To our astonishment, when we isolated the DNA from the T2 and T4 bacteriophages, we found that it could not be quantitatively digested to mononucleotides by the methods that had been successful with all other DNAs studied. It had been known that these particular viral DNAs had an unusual pyrimidine, 5-hydroxymethylcytosine, which replaced the usual cytosine. However, the resistance to the enzymatic digestion proved to be due to the presence of yet another, unsuspected, modification, a previously unknown pyrimidine. This was a glucose-substituted 5-hydroxymethylcytosine, which inhibited the enzymes. The extent of glucose substitution was 60 percent in T2 and 100 percent in T4; analysis of the progeny of phage crosses from mixed infections with T2 and T4 suggested interesting genetic interactions.

Many laboratories were working with T2 and T4 and these curious modifications of the viral DNAs spawned a variety of research projects, but I was soon diverted in other directions. As the research with φX progressed, it became more interesting and attracted an ever-increasing share of my attention.

After my experience at Caltech, I realized the importance of being informed of the most recent work in the field, which was now beginning to advance much more rapidly. Most of the research would of course ultimately appear in publications, but only after a delay of six months to a year. I had been a regular attendant at the summer Gordon Conferences; I now joined the American Society of Biological Chemists to be able to participate in its meetings, but these also were annual events. I maintained contact with colleagues I had met at Caltech. Also, by now some of our work had been published. This resulted in invitations to lecture at various universities and these visits broadened my network.

Through these activities, I became aware of the growing availability of support from the National Institutes of Health and the National Science Foundation. This possibility met a critical need. Having demonstrated that the RNA of the tobacco mosaic virus could be isolated as an intact molecule, and accepting the new dogma that nucleic acids were invariably the genetic material, I was led to the concept that the RNA of the virus might by itself be sufficient to induce infection, without any involvement of the protein component. Indeed, we were able to demonstrate infectivity with purified RNA preparations. However, per molecule, the RNA was about 1/1000 as infective as the intact virus. By chemical means, we could not exclude the possibility of residual protein equivalent to a 1 percent intact virus contamination of our RNA preparation. We needed a means to purify the RNA more completely.

This could be done in an ultracentrifuge, which could separate molecules by size. But we had none; indeed, there were none in the state of Iowa, and since they cost twenty-five thousand dollars, the campus could not provide one for us. I therefore applied to the NSF for an equipment grant, and in due time received the funds and installed the first ultracentrifuge in Iowa.

Then, just as we set out to use this instrument to provide data to verify our concept that the RNA itself was infective, a paper from the Virus Laboratory at Berkeley announced their results, indicating the infectivity of TMV RNA. Competition in the field, and from better-equipped laboratories, was clearly becoming more intense. We were not entirely satisfied with the quality of the evidence in the published paper and undertook a series of experiments that definitively correlated the infectivity in the RNA preparation with the sedimentation of the RNA in the ultracentrifuge. Nice, even elegant, work but in effect a gloss.

There had been only the most rudimentary studies of X-ray diffraction patterns from RNA. Having in hand this large, homogenous, biologically active RNA led us to undertake such studies. Of course, we hoped to find another striking regular structure such as the double helix. Unfortunately, as we now know, most RNA does not assume such structures. While we obtained some valuable information about nucleotide and internucleotide dimensions in the RNA, no dramatic structure resulted.

Because of the progress of our work and the gradual change in its emphasis toward more functional aspects, I began to attract graduate students from the biochemistry program, which was formally located within the Department of Chemistry. To facilitate this interaction, I was given a joint appointment with the Department of Chemistry. Having this dual affiliation had some advantage but, as I soon found out, also involved me in twice as many committee meetings and departmental disputes. But I enjoyed the personal contacts and could work within another domain to seek to increase the number of faculty doing work on the problems of what was coming to be called "molecular biology." I needed more colleagues, both for direct interaction and for their networks of professional acquaintance.

In the years 1953 to 1960, Max Perutz and John Kendrew develop the isomorphous replacement method of X-ray diffraction analysis and apply it to obtain the first detailed three-dimensional structures of the proteins myoglobin and hemoglobin. These structures incorporated the earlier models of Pauling and Corey but also provided specific details, down to atomic dimensions, essential to our understanding of the functions of the proteins.

As the genetic material, DNA must be accurately reproduced at each cell division. This reproduction is accomplished by a set of enzymes, DNA polymerases, which therefore play an essential role in heredity. In 1956 Arthur Kornberg isolates the first DNA polymerase, making it possible to study DNA synthesis outside a cell.

The hereditary information of DNA is contained in the sequence of its nucleotide subunits. This information is used in the cell to specify the sequences of amino acids in proteins. Translation from nucleotide sequence to amino acid sequence is accomplished in a complex chain of reactions, in which a key role is played by "transfer RNA" molecules. These molecules, with the help of enzymes, recognize both a nucleotide sequence and its cognate specific amino acid. Mahlon Hoagland and Paul Zamencik first describe transfer RNA in 1957.

12

Transition 3

At Ames in the early 1950s, I had the opportunity, indeed the luxury, to initiate and develop my research program at my own pace in relative isolation, drawing on my education and my imagination to explore an almost virgin field of great potential significance. In the expensive, intensely competitive, high pressure, teamwork world of molecular biology today, that would be impossible. Even by the mid-1950s, the field was attracting increasing attention and numbers, the pace of research was quickening, and isolation was no longer a boon but an increasing impediment.

In June of 1956, a major conference was held at Johns Hopkins University on the subject of "The Chemical Basis of Heredity." All of the leaders in the field were speakers and I was invited. At a break in an afternoon session, George Beadle steered me aside for a conversation. As was his custom, he came directly to the point. Would I be interested in moving from Ames to Caltech? I was pleased and surprised, but not totally. There had been subtle hints in recent correspondence with Delbrück, phrases to the effect that we might be seeing more of each other. My response to Beadle was yes, I would be interested, but my answer would of course depend on the specific conditions.

That fall I was invited to speak at Caltech at the dedication of their new Gordon Alles Laboratory. I spoke on "First Steps toward a Genetic Chemistry," reviewed the available knowledge of the chemistry of genes, and outlined the current problems and prospects for further advance. I also suggested that biology was entering a new era in which our science was at last beginning to acquire the breadth and depth of information requisite to a molecular understanding of complex biological processes. I used as an example the recent elaboration by X-ray diffraction of the atomic structure of the protein myoglobin by Perutz and Kendrew. The first analysis of such structures, this remarkable achievement in one step increased our knowledge about this protein by several orders of magnitude. I saw this as the harbinger of the influx of information that would surely, and necessarily, become available with newer techniques and instruments.

The talk was published as a lead article in *Science* and generated hundreds of requests for reprints. It was clear that this field was flourishing and was about to burst into scientific prominence.

Somewhat to my surprise, no further mention of Beadle's suggestion was made on this occasion and I returned to Ames rather perplexed. But not for long. Within a few weeks a letter arrived with a formal offer from Caltech. I responded with some questions about laboratory space and equipment and meanwhile, as was expected, informed my department chairman of this offer. The central administration of the campus had changed in the prior year. The new dean and provost called me in to discuss my situation. It seemed clear to me early on that they would like me to remain at Ames and would raise my salary to match, nearly, that which Caltech offered, but that they did not really expect I would decline Caltech. And they were right.

Academically and scientifically, the decision was easy. On a personal level, however, it was difficult. We had sunk roots in Ames. We had many close friends, our children had schoolmates. We had an established and increasingly active niche in the community. It was hard for me to break these bonds, harder still for my wife and children who could see little compensating benefit. Most distinguished scientists have moved, often several times, in their career. Such moves are most often professionally beneficial, but there is always a price to be paid. Deep friendships wane to acquaintance. Life becomes a discontinuous journey.

My acceptance of Caltech's offer was accompanied by some trepidation. I had proven myself at Iowa State. Was I really in the same class as the eminent scientists I knew to be at Caltech? Well, it seemed they thought so.

For reasons still obscure to me, after I had made my decision, I fell, for the first and only time in my life, into a depression. It lasted some two or three weeks. I knew at the time, each day, that it was abnormal, but I could not evade it. In this state, nothing seemed worthwhile. I went through the daily motions of existence, but life was totally flat and bleak. There was no joy, food had no taste, music did not please, the world seemed tinged with gray, all activities required effort beyond their worth. Worst was the sense, as day followed day, that this mood—unsought, undesired, unanticipated—might persist indefinitely.

I mention this because depression, like love and joy and anger and other emotional states, is only a word until it is experienced. Only with the experience can one understand and empathize.

As suddenly as it came, it lifted. The future—Caltech—was bright, truly bright, and I eagerly began plans for the transition.

13

φX

The genes were the goal and viruses were to be the key. Viruses bridge the boundary between the living and the nonliving. As do all life forms, they reproduce and undergo inheritable variation (mutation), but, lacking any metabolism, they can only do so within the confines of living cells.

Viruses were discovered toward the end of the nineteenth century as mysterious agents of infection. Invisible under an ordinary microscope, their features could not be discerned until the invention of the electron microscope. Many varied forms are known that can infect animal or plant or bacterial cells. Max Delbrück first recognized that viruses—in particular, bacteriophage, which infect the simpler bacterial cells—afford the potential for detailed analysis of the processes of reproduction and mutation.

Genes, known since the time of Mendel to be the determinants of hereditary traits, were for decades biochemically utterly obscure. Early on, they were thought, by a process of elimination, to most likely be protein. No other structures seemed sufficiently complex. As the true dimensions of the nucleic acids became more apparent, several lines of indirect evidence moved them, in some minds, into contention for the genetic role.

Whether proteins or nucleic acids, the properties of genes seemed nearly inexplicable. How could a complex substance, of whatever chemical form, be reproduced with high precision through thousands of cell generations? Even more strange was the fact that, when a gene was altered, mutated, by some unknown process, the mutant form was then reproduced just as faithfully as the original. And the mutant form could in turn be mutated further—or restored to its original state. Some, including Delbrück, proposed that a new principle of nature, a new physical law, might be needed to account for these phenomena. This speculation proved to be wrong—the ordinary principles of chemistry proved to be sufficient—but the proposal stimulated much discussion and fruitful research.

The choice of the bacteriophage-bacterium system proved to be most felicitous, providing a continual stream of new insights over a period of some forty years. In this period, the tools of molecular biology were developed and the study of bacterial virus infection provided a straightforward and rewarding field for their application and refinement. The simpler viruses proved to be only a coated set of specialized genes. By simultaneous infection of a population of bacteria, one could initiate within these cells a completely novel sequential pattern of genetic activity that could be dissected into discrete steps.

My research on the bacterial virus φX174 over a period of twenty-three years was a significant component of these advances, sometimes leading the stream, sometimes being borne by it. It was the centerpiece of my scientific career and is illustrative of the spectacular advances in our knowledge during this time.

Our aim was bold. We were to use this virus to pry open the processes of heredity and infection. We sought to find its genes, to count them, to deduce their mode of action, to understand their reproduction, to learn the basis for their stability and mutability, and to identify the agencies of their control. To know the genes of even one virus this well would be of profound importance in itself, as all life is related and the principles we found should be general. Knowing the genes would lead to understanding the nature of viral infection, to learning how the synthetic machinery of a cell is subverted to the production of more virus. And this might, in time, provide a basis for therapy of viral disease.

Initially, we had only the rudimentary knowledge that φX was probably small and could grow in certain strains of *E. coli*. At the end of our research, we knew the complete sequence of its DNA and the details of its genetic structure and had an extensive if incomplete picture of its architecture and the processes of its replication. This was a classic period for virology, especially bacterial, and research on φX was one of the central features. But the path was not straight, the way was often uncertain, and there were many surprises en route.

During my six months in Max Delbrück's laboratory in 1953, I had decided that while bacterial viruses provided in principle the simplest system in which to study DNA structure and expression, significant advance (using the techniques of that day) would require the use of the simplest of the bacterial viruses available. The T-even bacteriophages, then under the most intensive study, seemed much too complex to me. Proceeding on the (partially erroneous) assumption that the reproduction of smaller viruses would be simpler to analyze, I scoured the literature in search of small bacteriophages. Two—ϕX174, discovered in Paris, and S13, discovered in England—seemed suitable. The evidence as to their size was only qualitative and insecure, for each line of evidence could have alternative explanations. But these viruses seemed a good place to start. Remarkably, cultures of each of these were available in laboratories in France and England, and an available strain of *E. coli* proved to be a suitable host cell.

After my return to Ames, I began work on these viruses in 1953. Work with any new virus requires that it be "domesticated"—that one learn the better media in which to cultivate the host and the virus, the conditions suitable for its storage, how to purify it, and so on. In early experiments, ϕX proved hardier than S13, which lost viability rapidly on storage. From then on, we concentrated on ϕX though, as we later learned, these two viruses are in fact closely related.

After learning how to produce cultures of high infectivity, we increased the scale of culture to provide quantities (in milligrams) sufficient for purification. The most useful techniques involved various forms of centrifugation. After some stages of purification, we were able to associate infectivity with a particular component that, because it moved relatively slowly under centrifugal force, supported our hope that the virus would be small. I then examined, in the electron microscope, virus from preparations that appeared nearly pure in the ultracentrifuge. Small, approximately spherical particles about twenty-five to thirty nanometers in diameter appeared, providing evidence that the total mass could not exceed eight to ten million daltons.

All viruses, of the small number that had been analyzed, were known to contain a nucleic acid, either DNA or RNA. We assumed ϕX would too, but which would it be? No RNA bacteriophage were then known, but animal and plant viruses containing RNA were common. After trying several methods, we were able to separate the viral protein from its nucleic acid and demonstrate that the latter was DNA. But an oddity was immediately evident; its rate of movement under centrifugal force was such that were it a conventional double-stranded DNA, its molecular weight would exceed that of the entire virus. How could this anomaly be resolved?

The research to this point had required three years. I reported it at the ASBC meeting in April 1957.

Research was interrupted for a time by my move to Caltech in the summer of 1957, but I resumed the phage studies as soon as possible.

I set up a light-scattering apparatus at Caltech that permitted me to ascertain first the absolute molecular weight of the φX virus particle (6.2 million daltons) and then that of its DNA, which proved to be 1.7 million daltons. This molecular weight, together with its centrifugal and other properties, suggested strongly that the DNA could not be the usual stiff, double-helical form. Could it be a single-stranded DNA, then unknown in nature? Or more likely, was it a denatured, crumpled form of a double-stranded DNA in which the stiffening bonds between the strands had been disrupted somehow by the extraction procedure?

Several lines of evidence suggested that the state of DNA *inside* the virus particle was no different than that we observed after extraction. But the convincing fact proved to be the quantitative determination of the nucleotide composition of the DNA by the methods we had developed in our earlier work. This DNA did not have the Chargaff equalities of A = T and C = G. The DNA of φX had to be a single-stranded, completely novel form.

These results were published to some astonishment in the spring of 1959. In June of that year, in a lecture at the Brookhaven Symposium, I compared this discovery to that of "finding a unicorn in the ruminant section of the zoo."

In the six years after its formulation, the double helix structure of DNA had become dogma. It so neatly accommodated the known facts concerning DNA structure and replication that the proposed presence of a single-stranded DNA—in of all things a virus, a form specialized for reproduction—aroused amazement and doubt and stirred the imagination. Soon, many of the best graduate students at Caltech wanted to work in my laboratory, and I received numerous applications from postdoctoral fellows from around the world. Success breeds success, and the talent thus attracted surely facilitated further progress.

How could the single-stranded DNA of the φX virus reproduce? Was there another mechanism in addition to the mode of complementary replication implied by the Watson-Crick model of double-stranded DNA and demonstrated subsequently by the famous Meselson-Stahl experiment? To study this question, we undertook to follow the fate of the viral DNA after it was introduced into the bacterial cell. This was a novel, even daring, proposition that called on several of the newly developed techniques of molecular biology. Many of these techniques— the use of nonradioactive and radioactive isotopes, light scattering, the various forms of ultracentrifugation—were based on specific applications of physical principles. My background in biophysics served me well. The small size of the φX DNA, which permitted us to manipulate it with minimal degradation, was also critical.

We developed three distinct means to identify viral DNA within the infected cell and to determine its integrity. The most rigorous of these required measurement of the infectivity of naked viral DNA. According to the most common but not yet universally accepted theory, the protein coat of the virus merely served to protect the DNA and assure its entry into the infected cell. Once inside, the DNA, the genetic material alone, was thought probably sufficient to carry on the infection and give rise to progeny virus particles. If this were so, the free DNA would be infective if we introduced it into a bacterial cell.

We were able to accomplish this by preparing, under appropriate conditions, bacterial protoplasts—cells that lacked a layer of their cell wall and therefore were somewhat permeable to the free DNA alone. George Guthrie, then a graduate student, developed this technique and was able to achieve an infective efficiency of one infected protoplast per one hundred to one thousand viral DNA molecules. This figure was low compared to that of intact virus, but it was sufficient for our purpose. Theory aside, this was the first complete demonstration that a viral DNA was sufficient for infection. Combining all of the techniques, we could infect a cell and subsequently identify, track, and assay the infective potential of the intracellular DNA molecules of the parental virus. We could isolate these at various times after infection.

These combined experiments demonstrated conclusively that the parental viral φX DNA remained intact and infective on infection of the host cell. But within one to two minutes it was converted to a different form of DNA, which we called the "replicative form." Physical and chemical studies of the replicative-form DNA demonstrated that in fact it was the double-stranded version of the single-stranded viral DNA, paired with its newly made complementary strand. This replicative form was also infective to protoplasts. Further studies revealed that, after its initial formation, the replicative form reproduced as such, in the usual semiconservative manner of double-stranded DNA to produce some ten to twenty copies. This set of replicative-form DNA molecules then served as the template from which the distinctive single-stranded DNA of the progeny virus particles was made during the latter half of the infection. So replication of this single-stranded DNA in fact proceeded by the usual, complementary strand mechanism, albeit with some atypical aspects.

However, another major surprise was now in store for us. We had available from φX, for the first time, a homogeneous DNA of defined size (about fifty-four hundred nucleotides long) and nucleotide composition, with a specific biological activity (infection). It seemed a reasonable question to inquire whether there was anything special about the ends of this DNA. Was there a specific nucleotide or a specific sequence at either end of the chain? (Nucleic acid chains have a polarity and the two ends are distinguishable.) We had available two enzymes that we knew would degrade a polydeoxynucleotide chain from respectively one or the other end taking off one nucleotide at a time. Walter Fiers, a postdoctoral fellow, undertook to study the effects of these enzymes on the purified viral DNA and to characterize the products of the enzymatic action.

His results, repeated many times, were surprising. When he used highly purified enzymes, only a small fraction of the viral DNA appeared as mononucleotides, and these included all four possible in similar proportions. Further, there was no perceptible decline in the infectivity of the DNA after the enzyme digestion. Clearly, the bulk of the viral DNA, all of the infective DNA, was resistant to the action of these enzymes. Why? Were both ends of the DNA "blocked" by some special chemical group?

One day while walking back to the laboratory after lunch at the Athenaeum and pondering this question, I began to consider seriously a possibility we had occasionally tossed off in jest that the DNA might really be a ring—not a linear molecule, but a ring. A ring would of course have no ends to degrade, but if it were a ring it would behave slightly differently—it would move slightly faster under centrifugal force—than would the corresponding linear DNA of similar weight. Now in fact, in all of our DNA preparations, we had actually observed two components in the ultracentrifuge with closely similar rates of movement, present in varying amounts in different preparations. Preparations with the greater proportion of the faster-moving component had appeared to have higher specific infectivity. Our more recent preparations, made with better technique, had only small amounts of the slower component. We had had no explanation for this observation.

If the viral DNA were a ring, which would be the faster component, then the slower component *could* be rings that had accidentally been cleaved, linearized, at some random point and were therefore noninfective. Further, the ends of such randomly cleaved rings could account for the small yield of terminal nucleotides of all four kinds that we had observed on enzyme treatment.

An experiment to test this hypothesis became clear to me. I would start with a preparation of (almost) pure rings, then cleave them deliberately and slowly with an enzyme that would break the chains randomly. The proportion of rings cleaved could be followed by the loss of infectivity. As the rings were cleaved, DNA molecules would be transferred, first from the faster component to the slower component and then, as a second cleavage was made, out of the slower component into smaller fragments of random size.

The following afternoon I worked out the equations relating residual infectivity to the proportions of DNA in the first component, second component, and smaller fragments. Walter Fiers performed the experiment; it precisely confirmed the theoretical expectation. φX DNA had to be a ring.

I still consider this experiment, combining measurements of biological infectivity, ultracentrifugal analysis, and enzymatic activity, as probably the most elegant with which I have been involved.

The result, which I thought unarguable, met initially with considerable skepticism, particularly from nucleic acid biochemists. They were wedded to the concept of a polymer that, starting at one end, was extended unit by unit to the other. A ring molecule provided no purchase for such extension. If the single-stranded viral DNA was a ring, was the double-stranded replicative form also a ring? Alice Burton, then a postdoctoral fellow, carried out the analogous experiment, degrading the replicative form with an enzyme known to cleave both strands at the same site. She thus demonstrated its conversion to a linear, slower moving form in equal proportion with its loss of infectivity.

Then came an opportunity to satisfy the skeptics—seeing is believing. Dr. Albrecht Kleinfelter, then at UC Berkeley, had developed a method to visualize DNA in the electron microscope. It could only, however, be applied to double-stranded DNA. I took some replicative-form DNA to his laboratory, and we prepared the electron microscope grids and took pictures. When we examined the small pictures with a hand lens, the very first DNA we saw was a ring. Indeed, the field was filled with double-stranded DNA rings of the circumference to be expected for the known weight of DNA. The skeptics were convinced.

Later, when it became possible to visualize single-stranded DNA in the electron microscope, the rings of the viral DNA were quite obvious. Why had this result been so hard to deduce? In retrospect, we had been mesmerized by the predominantly linear character of DNA. To have a ring, one had to have an intact DNA. Prior to φX, everyone had studied fragments of DNA. Since then, many other ring DNAs have been found in viruses, plasmids, mitochondria, chloroplasts, and so on. In fact, it has become clear that for those DNAs that are not rings (as in eucaryotic chromosomes) special means must be provided to replicate their termini.

Now we had a complete DNA molecule, the complete genome of this virus. It was all there in our test tube. But how did it multiply? How did this relatively small molecule subvert the machinery of the infected cell to achieve its own reproduction?

Biochemically, the next logical step in our research would have been to determine the nucleotide sequence of the viral DNA, which could plausibly be expected to lead us to the genes through the genetic code then being deciphered. However, the technology to do this was not then available. We were able to separate out all the pyrimidine nucleotide tracts from the intervening purine nucleotide tracts and then to fractionate these according to size (thus learning that the longest consecutive run of pyrimidine nucleotides was eleven), but we could neither further fractionate nor sequence these tracts. So the information gain was limited.

We turned, therefore, to genetics. Using the recently developed methodologies for conditional lethal mutations (i.e., mutants able to grow under one set of conditions or in certain cells but unable to grow under other conditions or in other hosts), we were able to obtain a large number of mutants of φX. These could be classified into "complementation groups," or genes, by determining whether a successful infection would ensue in a cell infected with two different mutants. If so, then each mutant must supply the function defective in the other so that they were in different complementation groups. If such a joint infection were not successful, it was plausible that the mutants were from the same complementation group, that is, defective in the same gene. Nine complementation groups were discovered and named A–H and J. The mutations could be mapped—arranged in a linear order along a (necessarily circular) path—by the usual means of recombinant formation in genetic crosses. Only the mutants in genes D and E could not be consistently ordered for reasons that much later became apparent.

Studies of the abortive infective process in cells infected with these mutants provided clues as to the functions of the several genes. Gene E, for instance, was required for disruption of the cell membrane at the end of the infection, thereby releasing the progeny virus. In gene E mutants, the membrane failed to disrupt and viral synthesis continued extensively. As we could then artificially disrupt these cells, these mutants were useful to provide high yields of virus. Some genes proved necessary for early stages of DNA synthesis. Others were crucial for intermediate stages of viral particle assembly, while others coded for the four proteins of the viral coat.

Parallel with these studies, a long-term collaboration with Arthur Kornberg of Stanford proceeded intermittently. Professor Kornberg, a distinguished enzymologist, had devoted years to the biochemistry of DNA synthesis and had isolated and purified an enzyme (DNA polymerase) from the *E. coli* bacterium that could synthesize DNA using as a template whatever DNA it was given to copy. While chemical analysis indicated that the enzyme was making faithful copies, the ultimate test would be to demonstrate that the copy DNA had a true biological activity. With its defined size and a functional assay (infectivity), ϕX DNA was an obvious choice to test the faithfulness of Kornberg's enzyme. Could it make infective DNA copies in vitro?

Using our DNA, Kornberg's enzyme, and our infectivity assay, we tried in 1960. While considerable amounts of DNA were made in excess of that initially added, infectivity fell rapidly. Attributing this result to impurities (degradative enzymes) in the DNA polymerase preparation, we tried the experiment again in 1963 with a more highly purified enzyme. The same result.

With the discovery that ϕX DNA was a ring, it became evident that effective synthesis would require an agent to close the newly made DNA strands into a continuous circle. A recently discovered enzyme, DNA ligase, had the ability to do this.

In 1967, with the help of Mehran Goulian, a postdoctoral fellow in Kornberg's laboratory, we set out again to make infective DNA using DNA polymerase together with DNA ligase and ϕX DNA. To insure that the synthesized (and, we hoped, infective) DNA was distinct from the input viral DNA, we incorporated a density label so that the product could be physically separated from the template. The experiment was eminently successful. Together, we had made for the first time an infective agent in the test tube capable of thereafter reproducing itself indefinitely in host cells.

In truth, we had only copied faithfully an existent design. But our copies were, in fact, potentially immortal.

We reported this work in the December 1967 issue of the *Proceedings of the National Academy of Sciences*. By some circumstance, this report landed on the desk of an aide to President Johnson on the day the president was to speak at the National Institutes of Health. It was incorporated into the president's speech as "the nearest approach to the synthesis of life in a test tube," an instance of the remarkable research supported by NIH. As a consequence, this result was reported in newspapers and news magazines everywhere. A result of such coverage was the receipt of many letters, some from long lost friends or relations, some quite touching. The newspaper articles, referring to the small size of φX, sometimes called it a "dwarf virus," and I received several letters from people suffering from dwarfism who asked if my research could alleviate their affliction. Sadly, not.

In the early 1970s, our ability to analyze DNA was greatly augmented by two developments. The restriction enzymes were discovered, which cleaved DNA only at specific nucleotide sequences. Using these, a DNA such as that of φX could be broken reproducibly into a small number of specific, isolatable regions. By the use of partial digests, Amy Lee in my laboratory was able to order these regions into a continuous map, the first complete restriction fragment map of a DNA. By hybridizing one or more of these fragments with the viral DNA, Lee Compton was able to visualize specific regions of phage DNA in the electron microscope.

Even more significant, Fred Sanger at Cambridge University was developing means to sequence the DNA of restriction fragments nucleotide by nucleotide. As a defined DNA of specific size with an ordered restriction fragment map, φX was the natural choice for Sanger's first application of this technique. After obtaining his Ph.D. in my group, Clyde Hutchison went to Sanger's laboratory to help with this project. In 1977, Sanger announced the complete nucleotide sequence of φX, 5,386 nucleotides. This was the first complete sequence of any DNA, a major accomplishment. As such, it marked the end of an era of φX research and the beginning of a vast new period of DNA research.

Sanger's sequence indicated clearly the location of the genes we had laboriously sited on the genetic map. It also contained one major surprise. We had been unable to map genes D and E in a consistent manner; the reason now became clear. The two genes used in large part the same region of DNA, reading it in different, overlapping trinucleotide frames!

This possibility had been discussed before but was usually dismissed with the argument that such an overlapping pattern would result in such great constraints of the choice of amino acids that it seemed very unlikely that two functional proteins could result. Yet here it was!

In small viruses such as φX, the physical constraint on the size of the DNA evidently drives the virus in an evolutionary sense to make the very maximum use of its fifty-four hundred nucleotides. Other small overlaps, at the ending of one gene and the start of the next, attest to the same pressure, with the use of one region of DNA to code for portions of two proteins.

At this same time, φX was also serving an important role in the attempt in Arthur Kornberg's laboratory to reconstruct from *E. coli* extracts an entire system able to initiate, extend, and complete DNA synthesis. The use of φX DNA as a small, defined template enabled this laboratory to extract and successfully refine the many factors needed, in addition to the DNA polymerase and ligase, to reproduce DNA in a truly biological fashion. Out of this work has come the recognition that while the smaller viruses, with less genetic material, contribute fewer components to the infective process, the host cell must therefore contribute more. In brief, the smaller viruses are the more parasitic; their infective process, resulting in viral reproduction, may not be appreciably simpler than for larger viruses.

While certain aspects of φX replication, in particular the manner of particle assembly, remained obscure, by 1977 φX had fulfilled its primary role in the advance of molecular biology. My original conception in undertaking φX studies—the need for a simple, defined, biologically assayable DNA as an experimental subject—had been validated. The research with φX had played a significant role in a classic period of advance in our understanding of genes. The technology was now available to attack more complex problems.

This tale—the unraveling and elucidation of the structure and process of replication of one virus strain—can serve as a prototype for much of biological research. Starting with but a few empirical observations, each finding led to deeper questions of detail or process. To answer these questions required the development, application, and refinement of new techniques and new approaches. Some came from our laboratory; others were borrowed from Kornberg's laboratory, or Sanger's, or were culled from the literature on restriction enzymes, conditional lethal mutations, or ultracentrifugation. Out of this study of one virus came often surprising discoveries of great generality—that there are viruses containing single-stranded DNA; that many DNAs are rings; that single-stranded DNAs reproduce via the conventional complementary mechanisms; that, in small viruses, the same DNA region can be used to determine more than one protein—as well as the important demonstrations that an infective DNA can be synthesized in vitro and that a complete genome can be sequenced.

These are the rewards of a sustained, collective effort: meaningful contributions to human knowledge. They were not easy won—one does not write of the failed experiments, the misconceived, the accidents, the equipment failures, the blunders. One cherishes the nuggets of knowledge, and the students who mastering new skills on these problems have gone on to make advances in many other fields.

14

The Caltech Years—
The Fulfillment

My first decade at Caltech was exhilarating. Not only was the φX research progressing steadily and successfully into previously inaccessible processes, but more broadly it was a breakthrough era for molecular biology. The many ideas, hypotheses, and concepts immanent in the double helix model for DNA were about to receive experimental test and verification or refutation. Within the next half dozen years the "semiconservative" model for DNA replication was demonstrated, the genetic code was deciphered, and "messenger RNA" was discovered as the intermediate in the conversion of genetic information into protein structure. Caltech played a significant role in each of these advances.

The invention and theoretical analysis of the technique of density gradient centrifugation by Jerome Vinograd at Caltech provided the methodology for the brilliant experiments of Meselson and Stahl, as well as some of our own crucial φX experiments. Using the stable heavy isotope of nitrogen (N^{15}) as a density label, Meselson and Stahl were able to demonstrate that, on replication of DNA, the two strands of the double helix dissociate; each strand, remaining intact, becomes one of the two strands of each daughter double helix. This process had been postulated but, as we were completely ignorant of the complex enzymatic processes involved, had remained in doubt until experimentally verified.

The correspondence between DNA as gene and protein as "gene product" was clear, but the nature of the biochemical connection was unknown. Progress in the achievement of protein synthesis with extracts from disrupted cells had indicated that proteins were actually synthe-sized on subcellular particles called ribosomes. But the ribosomes of a cell were chemically uniform. How then were individual ribosomes *pro-grammed* to produce different proteins? Experiments in several labo-ratories, including particularly research conducted by Brenner, Jacob, and Meselson at Caltech, demonstrated that a minor, ephemeral variety of ribonucleic acid, called "messenger RNA," was made by copying into RNA portions of the cellular DNA. Such messenger RNA mole-cules then served for a short time as programs to be read by the ribo-somes, to produce the specific proteins.

Clearly, the genetic information contained in the sequences of nu-cleotides in DNA, as copied into the messenger RNA, somehow spec-ified the sequences of amino acids in the various proteins. But the code connecting these two sequences was unknown. Delbrück was very in-terested in this and there was much speculation and many seminars about the nature of the genetic code: overlapping codes, self-correcting codes, commaless codes, two-letter, three-letter, four-letter, six-letter codes were discussed with proponents of one or another variety pro-claiming its apparent or unique merit. It was a game many could play and it seemed that if one were really clever, one could deduce the one clearly optimal code that nature had chosen to use.

But the experimental problem of deciphering the actual code seemed difficult to resolve until the highly fortuitous discovery by Marshall Ni-renberg that, under nonphysiological conditions, protein-synthesizing systems extracted from cells could use artificial polynucleotides as mes-senger RNAs. Such synthetic polynucleotides could be made of very simple sequences, resulting in the synthesis of correspondingly simple chains of amino acids. The obvious correlation then of amino acid se-quence with nucleotide sequence permitted the complete decipherment of the genetic code within a few years.

At this same time, the discovery of the "amber" viral mutants by Richard Epstein at Caltech—mutants that could only grow in cells that used a slightly altered genetic code—opened a wide new path to the identification of specific genes and the analysis of their function.

It was an heroic time in which individual scientists working with small groups could, with great effort and ingenuity, make momentous discoveries. These discoveries, over a space of little more than a dozen years, transformed biology by exposing the molecular machinery of ge-netic reference, thereby prying open and making accessible the basic processes of life.

It is hard now to realize how difficult those first steps were, how months of step-by-step experimentation and correction were required to obtain a result that could now be achieved in a few hours. But it was accomplished, and those involved will always remember the thrill as each new discovery was announced, each new perspective revealed.

After these discoveries of the 1950 to 1965 period had been consolidated, some, notably Gunther Stent, proclaimed the end of the "golden age" of molecular biology: all of the great discoveries had been made and, most disappointingly, no new fundamental principles had been discovered; biology turned out to be a specialized form of chemistry. But Stent's vision was woefully short. He did not foresee the second and even greater flowering of molecular biology.

The invention of recombinant DNA in the early 1970s by Herbert Boyer and Seymour Cohen enabled the extension of molecular biology techniques far beyond the small discrete nucleic acids of viruses to the much more complex nucleic acids of higher organisms. And the development of methods for the determination of nucleotide sequence in DNA by Walter Gilbert and Fred Sanger improved the effective discrimination of these techniques by orders of magnitude, permitting the detailed analysis of mutation and genetic control mechanisms. The previously inaccessible fields of developmental biology and physiological control then became open to analysis at the molecular level.

In this same period, Caltech was the scene of other remarkable scientific accomplishments. Martin Schmidt and Jesse Greenstein discovered quasars, the intensely powerful and distant sources of radiation that are believed to be at the centers of galaxies. Murray Gell-Mann developed his quark model of the structure of subatomic particles, which brought coherence to an increasingly diverse and puzzling set of data. The Jet Propulsion Laboratory, after some difficulties, successfully sent the first probes to land on the moon.

Richard Feynman in physics was the one true genius I have known. He had an unsurpassed depth of understanding of the fundamental principles of physics in the broadest sense. I watched in astonishment as, on more than occasion, he swiftly applied this deep insight to completely novel situations and assorted and arrayed the important variables in some comprehensible manner.

What special qualities made Caltech the locus of many of these advances? Caltech is unique. A small, private school, focused on basic science and advanced engineering, it attracts the brightest freshmen (as measured by SAT scores) of any college or university in the country. Because the freshmen class is limited to 220 students, now about 25 percent female, the resultant undergraduate student/faculty ratio of three to one permits individual student recognition. At the same time, the faculty is able to devote most of its time to research, working with some one thousand graduate students and four hundred postdoctoral fellows. The concentration of interests, the uniformly high level of talent, and the opportunity for close interaction with faculty make it an excellent school for students who *know* their educational goals. For others its opportunities can be too limited.

The renown of the institute has enabled it to develop extensive funding and it has the highest ratio of endowment per faculty member of any private university. Its ability to provide substantial research resources in support of individual faculty members has in turn enhanced the ability of those members to achieve and maintain eminence in their fields.

Absent the elemental necessity to hire enough faculty to provide the needed classes for large numbers of students, Caltech can be highly selective in its choice of permanent faculty. Similarly, as a private institution, Caltech can be highly selective about the fields of education and research it chooses to undertake. This freedom to focus energy and resources on a few specific areas within, for instance, biology or physics has permitted the institute to deliberately select the most promising frontiers in any period and to develop a "critical mass" of interactive faculty to explore and exploit those fields.

Careful screening among the most promising applicants for junior faculty positions is followed later by application of rigorous criteria for the award of tenure. As a general principle, tenure is awarded only to junior faculty members who have demonstrated accomplishments sufficient to place them among the "top five" in their field and age group in the country. In consequence, 21 percent of its faculty are members of the National Academy of Sciences and 8.5 percent are members of the National Academy of Engineering.

The small size of the Caltech faculty facilitates scientific contacts among the several disciplines. Such interaction is fostered by the presence of a rather elegant faculty club, the Athenaeum.

The Athenaeum often provided an intellectual feast, a gourmet meal for a science junkie. One might talk with a geologist about the age of the earth, the composition of moon rocks, or the temperatures of ancient seas. Or with an astronomer about the precise periods of pulsars or the mysteries of quasars. Or with a physicist about the zoo of fundamental particles, the oddities of quarks, or the postulated gravity waves. Or with an engineer about supersonic planes and rockets, space exploration, lasers, or the potential of computers. Or with economists or historians to hear their perspectives on the events of the time. To be sure, there was always politics or the fortunes of the Dodgers to fill in the gaps.

As compared to MIT, Caltech is much smaller, with about one fourth the student body. Caltech is much less diverse, with a narrower spectrum of engineering disciplines and no schools of architecture or management. Basic science plays a larger and engineering a lesser role. The breadth and scale of MIT permit it to consider problems of large societal concern and impact such as transportation, urban planning, and industrial competitiveness.

Caltech is quite comfortable with itself in its present format. It has a self-image of singular superiority, which is not quite as readily justified as formerly. For, while still unique, Caltech is not as peculiarly distinctive as it was in an earlier era. Other, much larger institutions have developed concentrations of faculty in one or more specific areas able to rival those at Caltech. But it remains an extraordinary constellation of excellence in its chosen sectors. While the limitations imposed by its narrow focus on curricular offerings and research opportunities are recognized by some, the institute has not been sufficiently motivated to undertake the major effort required to transcend them.

When we arrived in the 1950s, Pasadena still had a small-town atmosphere. Unfortunately, at times it also had intense smog generated in downtown Los Angeles and blown up against the mountains. Nevertheless, there were pleasant residential neighborhoods within walking distance of the Caltech campus in which many faculty lived. We found a residence nearby that greatly facilitated return in the evening to initiate or complete laboratory experiments. This ease of access also facilitated attendance at evening lectures and engendered a sense of campus community.

The biology faculty at Caltech was, and continues to be, extraordinary. When I arrived, George Beadle was the chairman and unquestioned leader, a "father figure" to most of the department. When he subsequently became dean of the faculty and proposed that the department find another chairman because of the demands on his time, we would have none of it. Only when he left to become president of the University of Chicago did we replace him.

Henry Sturtevant was the "grand old man" of the genetic community. A student with Morgan in the early days of drosophila research at Columbia, Sturtevant had known personally almost all of the great figures in the history of genetics. He had an encyclopedic knowledge of genetics and entomology as well. As I saw him age and reach retirement, I realized how much is lost when such an irreplaceable mind is gone. Ed Lewis, his protégé, would later carry on the drosophila tradition with great distinction. Sterling Emerson, Ray Owen, Norman Horowitz, and Herschel Mitchell completed an extraordinary grouping of geneticists.

Frits Went with his Phytotron, in which he could grow plants under precisely controlled conditions, and James Bonner and Arie Haagen-Smit, who were more biochemically oriented, constituted an impressive program in basic plant biology. Haagen-Smit had already become interested in the chemistry of the pervasive smog problem and had deduced the primary chemical reactions involved. Meanwhile, Henry Borsook, a long-time biochemist, was pursuing protein biosynthesis, a problem that would only be solved subsequently by much newer approaches. "Case" Wiersma and Anthonie Van Harreveld were classical neurophysiologists engrossed in electrophysiology, while Roger Sperry undertook his brilliant and idiosyncratic studies of the factors determining neuronal connections.

All told, a remarkable group.

Caltech provided me the opportunity for research at a higher level of intensity and on a different scale than had Ames. Very able graduate students wanted to work in my laboratory; postdoctoral fellows from Europe, Asia, and the U.S., eager to have experience at Caltech, soon applied to study with me. Over the years, I had scholars from Great Britain, the Netherlands, Belgium, France, Sweden, Germany, Czechoslovakia, Hungary, the Soviet Union, Japan, and Chile work in my laboratory. It is a great pleasure now to be able to travel about the world and visit former students and see how well they are doing in their science in their native countries. Science is truly an international activity. I can walk into a lecture room in Kyoto, Santiago, or Stockholm and see the periodic table on the wall and feel at home.

And, as well, I regularly had one or two bright undergraduates, eager to begin research, working part-time with me.

Teaching responsibilities at Caltech were much different than at Iowa State. Formal classroom obligations were much reduced and there were no large "service" courses akin to engineering physics. However, there was much greater informal educational activity, of a more personal and focused nature, with graduate students and postdoctoral fellows in my laboratory and in small seminar-type courses in which a current topic of interest would be discussed in depth.

"Introduction to Biology," which was taken for one academic quarter by about half of the freshman class, was the largest biology course; I taught this for six years and enjoyed presenting the essence of the new biology to such bright students. In later years, I introduced a course in "Biology and Society" that considered the social and ethical issues arising from the advances in biology and medicine.

In addition to attendance at seminars and lectures, I sought to broaden my knowledge of biology by sitting in on various classes. Max Delbrück presented a one-quarter lecture course each year on a topic of his choice. He took this quite seriously and used the course as an opportunity to learn about a subject of interest. One year he talked about biological membrane structure and function, another about basic solid-state physics and order-disorder phenomena, a third about transducers in sensory processes with particular discussion of the optic retina.

Caltech undergraduates are remarkable. Uniformly bright, with high analytical skills, they have a focused intensity that certainly surpasses the students of my MIT days. Perhaps because they are more oriented toward science than engineering, the student culture is less attuned to the larger world and thus more susceptible to the fascination of research and scientific discovery.

As one result of the exclusively technical emphasis at Caltech, life for students and faculty can be remarkably insulated from even major external political or social concerns. In the 1960s we observed the upheavals, riots, and drug scenes on such campuses as Berkeley and Columbia from a distance with dismay and some disdain. Caltech students were too serious and committed for such folly. But our own children were not immune to the peer movements of the time and through them we achieved at least some understanding of the youthful idealism and depth of feelings involved. And Vietnam reached even Caltech, however peacefully, with large discussions and teach-ins.

From the beginning, this military action had made no sense to me. Vietnam seemed to be primarily a civil war with which the U.S. had no great concern. And as the war dragged on, with growing casualties and an evident unwillingness in Washington to do what would be necessary to win it, our involvement seemed ever more a cruel and grotesque mistake—one from which the nation would long suffer.

A senior faculty position at Caltech conveys a prestige that launches one into larger spheres of activity. Soon after I joined the institute, John Kendrew came through to see me about starting a new journal. Publication in the field of molecular biology was still hindered by the too-specific interests of existing journals. Out of this came the *Journal of Molecular Biology* with Kendrew as editor-in-chief and myself as one of the associate editors. It is today still a leading journal in the field of macromolecular structure and function. I served as an associate editor for ten years. After this, I served for five years as a member of the editorial board of the *Annual Reviews of Biochemistry,* and, subsequent to that, for eight years as chairman of the board of editors of the *Proceedings of the National Academy of Sciences.* Such service is important to the maintenance of the scientific standards of publication in the field; it also enables, indeed requires, one to keep up with the advances across a much wider sector of biology than one's own research interest. However, it requires a considerable investment of time and intellectual energy that is not usually professionally rewarded.

I was asked to chair—which meant to draw up the program and select the participants—the 1960 Gordon Conference on Proteins and Nucleic Acids. This was the last year when both fields could be considered together. Organizing such a conference is a considerable logistic task and the supporting services of an institution such as Caltech are essential. It is a pity that video records could not have been made of these Gordon Research Conferences in the 1950s and 1960s. It was here that great advances and important smaller steps were often first presented, discussed, and critiqued in the informal, open, and friendly fashion of science.

Here Fred Sanger presented the first amino acid sequence of a protein, insulin. The sequence of the adrenocorticotropic hormone was not far behind.

In the early days, DNA or RNA preparations from cells were, of necessity, treated as homogeneous substances. While, for many purposes, this proved to be valid for DNA, many years and many conferences were required to sort RNA into its various classes of messenger, transfer, and ribosomal RNA with different structures, functions, cellular locations, and turnover rates.

At the Gordon Conference, Julius Marmur and Paul Doty described their remarkable discovery of DNA renaturation, the basis of much of today's molecular genetics. The enzymology of DNA and RNA was progressively elaborated. Here, too, Francis Crick presented his "adaptor" hypothesis to link nucleic acid sequence to amino acid sequence and his "commaless" code. Mahlon Hoagland and Paul Zamencik described their discovery of transfer RNA. Bacterial transforming factors and viral structures were analyzed and the relationship of DNA structures to genetic factors was progressively refined. Gradually, year by year, the modes of synthesis of DNA and RNA and protein were clarified.

I will never forget Seymour Benzer's succinct interruption of a prolonged presentation by Gunther Stent. Gunther, in a late evening session, was describing in endless detail a complex bacteriophage experiment that had yielded negative results. Finally Seymour spoke up, succinctly and loudly: "Big deal!" The meeting abruptly adjourned to the bar.

Each year the field progressed and expanded with new knowledge and new researchers. After 1960, the meetings had to become more numerous and more specialized.

In 1960 I received my first invitation to an international scientific conference at Chamonix in France and so had my first experience abroad. Europe was a daily surprise. We landed in Copenhagen and I was stunned to be in the midst of crowds of people all speaking a tongue that meant nothing to me. *Of course*—but it felt like a verbal assault. In Europe, at least I could read the signs in the streets and at terminals. A few years later in Japan, I realized what it must be like to be illiterate, to be totally unable to find one's way—or worse, even to ask.

We visited the major cities: Copenhagen, Brussels, Paris, Geneva, London. The wealth of architecture and art accumulated over the centuries was astonishing. London still bore the scars of the wartime bombing. Each distinct sector of Paris entranced us. I was struck by the omnipresent weight and legacy of the past, so different from the ever-changing life at home.

The remarkable progress in molecular biology received wide attention. Many universities had not kept up and now lacked faculty in this suddenly important area of biology. I began to receive unsolicited invitations from other universities, many quite distinguished, to join them, often accompanied by commitments to develop a major program in molecular biology that I would head.

At first I was flattered and considered a few of these sufficiently seriously to visit and learn of their plans and evaluate the opportunity. But I soon realized that I simply did not want to leave Caltech to be a professor elsewhere. I had all of the resources and opportunity I needed, I had first-rate colleagues, and I liked the living arrangements. Thereafter, although I continued to receive such invitations, I simply declined with thanks, not wanting either to waste their time or to stir up uncertainty in my own life. In a few instances, however, I fear my refusal even to look at their offer was misinterpreted as arrogance.

According to Watson-Crick double helix model for DNA, the two strands of the double helix should be separated upon replication of the DNA. In 1958, in an important test of the model, Matthew Meselson and Frank Stahl experimentally confirm this separation.

Following the discovery of the double helix structure, it was expected that all DNA would be double-stranded. It is thus a considerable surprise when, in 1959, the DNA of the virus φX is shown to be single-stranded.

In 1960, Paul Doty and Julius Marmur discover DNA renaturation, wherein if the two strands of a DNA double helix are separated in solution, they will subsequently pair up in precise register and reform the double helix. This process forms the basis for much of the technology of molecular genetics.

Also in 1960, Marshall Nirenberg produces, in vitro, a polypeptide translated by a cellular extract from a synthetic polyribonucleotide. Further exploitation of this discovery by Nirenberg, Severo Ochoa, and Gobind Khorana leads to the complete elucidation of the genetic code linking nucleotide sequence to amino acid sequence.

DNA does not serve directly as the source of information for protein structures, but is first copied into another form of nucleic acid, called "messenger RNA." Messenger RNA then participates directly in protein synthesis. In 1961, Sydney Brenner, François Jacob, and Matthew Meselson first describe messenger RNA.

DNA was thought to exist exclusively as long linear strands. In 1962, the DNA of the virus φX is demonstrated instead to be a ring molecule, a structure since shown in many viral and other DNAs.

In 1967, Mehran Goulian, Arthur Kornberg, and Robert Sinsheimer reproduce an infective DNA (φX) in vitro. This accomplishment confirms the validity of many of the biochemical procedures then in use and opens the door to the test-tube synthesis of infective DNAs.

Research upon DNA was long inhibited by the inability of scientists to break the long chains into defined, reproducible segments. In 1970, Hamilton Smith first characterizes an enzyme, a "restriction endonuclease," that can cleave DNA only at specific nucleotide sequences and thereby reduce a chain to a specific set of fragments. Hundreds of such enzymes, with varied sequence specificities, are subsequently isolated and form an indispensable tool for molecular genetics.

In 1973, Stanley Cohen and Herbert Boyer invent plasmid cloning, thereby making recombinant DNA a practical tool. Plasmid cloning permits the amplification a millionfold or more of any specific piece of DNA isolated by the use of restriction enzymes or synthesized in vitro.

In 1977, Fred Sanger describes the first complete sequence of a DNA virus, ϕX. Knowledge of this complete sequence, 5,386 nucleotides in all, permits the definitive determination of all of the genes of this small virus and provides the first instance of overlapping gene sequences.

15

Of Honors, Awards, and Prizes

On a May day in 1967, I received with a mixture of surprise, pleasure, and puzzlement a letter informing me that I had been elected a member of the National Academy of Sciences. Surprise because I had not known that I was being considered; pleasure because I was aware, though not very, that the academy was a prestigious organization and membership was considered an accolade; and puzzlement because, although I had been a scientist for over twenty years, I knew almost nothing about this organization, about how one became a member, or even whether I wanted to be one, although now I clearly was, though I had never been asked.

As I subsequently learned, the National Academy of Sciences is an extraordinary organization, unusual in almost every feature. It is a quasi-governmental organization, founded during the Civil War with a congressional charter to make available to the federal government the expertise of the scientific community. Yet it is also, and importantly, a quasi-private organization that receives no direct funding from the government, that elects its own members and officers, and that can initiate studies and projects of its own in addition to those it undertakes on request for the government. Located in Washington, the academy serves in some degree as an intermediary between the intensely political world of the Capitol and the academic, sometimes arcane world of the scientist.

The role of science has changed drastically since the Civil War. Government has expanded and most government agencies have acquired their own in-house technical expertise. At the same time, government has become the principal patron of science and federal funding is the sine qua non of research in every field. The government uses the academy as a source of external, impartial, expert advice on a host of technical issues arising in modern society. The academy is able to provide this because of the scientific eminence of its limited membership and because of a conscious effort made to establish balanced panels to analyze each issue, followed by scrupulous, thorough review of reports before they are released.

While the academy thus serves an important national function, its procedures and activities are in many ways baroque and it has some major limitations.

I was elected to and served on the Council, the nominal governing body of the academy for three years, 1970 to 1973. The academy also sponsors a major scientific publication, *The Proceedings of the National Academy of Sciences,* and I served as chairman of its board of editors for eight years, from 1972 to 1980, all of which both permitted and required me to achieve some understanding of the capabilities and limitations of the academy.

A remarkable amount of the time and energy of the academy is devoted to self-perpetuation, to the election of the new members. Only sixty may be elected in any year and while this number has fluctuated somewhat over the years, it has in no way kept pace with the increase in the number of active scientists, so membership has become progressively more exclusive. Since new members are nominated only by present members and are elected by vote of all members, not surprisingly membership tends to be concentrated in a small number of elite research universities—Harvard, Berkeley, MIT, Caltech—and in certain disciplines. The latter effect, which militates against election of members in newer disciplines, is from time to time countered by action of the academy's Council, which can provide for special nominations that bypass some of the lower-level hurdles to election. The former effect, institutional concentration, is never addressed and indeed is considered reasonable and proper.

Though the academy was initially dominated by the physical scientists, in more recent years the biological scientists have achieved nearly equal status. The divisions of the academy representing applied science, the medical sciences, and the social sciences have distinctly lesser influence. In fact, the limited role accorded to the applied sciences in the academy inspired the formation in the 1970s of a parallel sister organization, the National Academy of Engineering, established by engineers who felt that their profession was not receiving appropriate credit for its national contributions. The two separate academies collaborate in certain responsibilities and coexist in a sometimes uneasy partnership.

In an effort to stave off the formation of yet another academy, the National Academy of Sciences established a somewhat autonomous branch, the Institute of Medicine. The institute is concerned primarily with the sociological, organizational, and policy aspects of medicine as a profession, as distinct from its scientific component, which is left primarily to the academy.

As membership in the academy is a recognition of distinguished scientific contribution, it tends to be a venerable institution, with the average age of election in the fifties. After I was elected, Ray Owen remarked that he particularly enjoyed the academy meeting as one of the few where he could still be among the younger group present. However, while advanced in years, the membership is by no means devoid of spirit, as was evidenced by the sight of several hundred academicians marching in the aisles while the Preservation Hall Jazz Band played "When the Saints Come Marching In" at an evening concert.

While the Council, twelve members elected for staggered three-year terms, is the nominal governing body, the president is the only full-time officer, elected for a six-year term and eligible for reelection. The president has a staff and, as he or she is present before, during, and after the monthly Council meetings, plays a major administrative and policy-making role. During my period of major interaction with the academy, Philip Handler was president. Aristocratic, eloquent, skilled in the ways of Washington politics, he ran the academy in an increasingly personal fashion as his twelve-year term progressed. The membership was by and large satisfied with the arrangement as long as it determined those features in which it had real interest, the disciplinary organization of the academy and the distribution and election of new members.

As one who had lived within the cloistered walls of academe, I at first found the wholly political life of Washington—in which who one knows, and how well one knows them, can be far more important than the logic of an issue—bizarre and uncongenial. As I became more familiar with this scene and realized its imperative, I became more appreciative of the role that Handler and those like him played in meshing the needs of science with the wishes of Congress and the administration.

The academy is able to serve the government well in its basic role as the provider of dispassionate advice on technical issues. The academy cannot serve the government as provider of advice concerning science policy. Issues such as the allocation of resources among disciplines or the establishment of priorities, which thrust different factions of the academy into competition, cannot be resolved within the academy. Nor has the academy been notably successful in the formulation of policy with respect to those social issues for which the scientific and technological factors are significant but not overriding components. While one might hope for a more rational approach to the resolution of such questions, in practice the intricate political compromises of Washington appear to generate the only acceptable solutions.

As chairman of the board of editors of *The Proceedings*, I was responsible for maintaining the quality of its articles, assuring its fiscal solvency, and placating occasional irate members. While the content of *The Proceedings* ought to reflect the diverse interests of the academy membership, in practice the publication tends to be used principally by a selected subset. In recent decades, papers of the biologists—particularly the molecular biologists—have dominated the pages of *The Proceedings*. Of course, all scientific journals are subject to a specialization of readership and authorship. Authors want to publish in those journals that are read by the prominent workers in their field. Repeated efforts to diversify the scientific fare and stimulate the submission for publication of papers in the physical or social sciences have failed. These disciplines have their own journals and it is in those journals that readers look to learn what is new. A new development in physics published in *The Proceedings* would not be seen by most physicists because *The Proceedings* has little of interest to them.

Members have the privilege of submitting self-authored articles to *The Proceedings* without review. The assumption is that articles from such distinguished scientists will surely be meritorious. Mostly this is true, but occasionally, especially in articles submitted by some of the more senior members, this standard is not met. The chairman of the board of editors has the authority to reject a paper even when submitted by a member. If the paper was outright embarrassing or, say, bore on medical concerns and—given the prestige of the academy—might induce some patients to forsake other more efficacious treatments, I would write to the author to suggest that the paper was not up to his or her usual standard. Outright rejection never proved necessary.

And I will always remember the fine paper submitted by Joel Hildebrandt at the age of ninety-six, together with a handwritten cover letter in a perfect, unfaltering hand.

For the most part, scientists do not become wealthy. In some disciplines, industrial consulting and entrepreneurial opportunities provide exceptions, and a few have written lucrative textbooks, but in general scientists are far less well remunerated than are other professionals such as physicians and lawyers, not to mention MBAs.

As one consequence, alternative modes of reward—awards and prizes—have been developed. These are primarily honorific, providing only modest financial sums. Such modes of recognition are usually found in the more mature disciplines and are sponsored or dispensed by well-established professional societies. The awards are often named for a distinguished member, usually deceased, of the discipline, and the selections are made by senior members of the society.

In general, these prizes recognize a distinguished career of significant contribution in a particular field and as such are awarded to scientists at the peak of their productivity, or beyond. Some awards are specifically designated for younger members of the profession. Receipt of such an award can have a favorable effect on one's career. In a sense, such an award is a self-fulfilling prophecy, for the recipient is more likely to obtain a better academic position, to attract better students, to receive larger grants, and so on. In science as in other aspects of life, there is a natural tendency for "the rich to get richer." Regrettably, such awards are not usually available in the newer interdisciplinary or ground-breaking fields, which often do not fall within established professional boundaries. Their lack expresses and reinforces the incrementalism and rigidity of the disciplines.

I have been honored by two major awards; both came a few years after the most significant ϕX discoveries.

In 1968, I received the California Scientist of the Year Award. Unlike most, this is a cross-disciplinary award given to the California scientist in any field whose accomplishments in the previous year are deemed to be of the greatest significance. The award is sponsored by the Museum of Science and Industry and the selection made by a panel of senior California scientists. Luis Alvarez was chairman of the panel that awarded my prize. In addition to the citation, presented at a black-tie banquet attended by distinguished members of the community, a five thousand dollar prize was also awarded. Receipt of this award made me aware of the incidental benefits of such an accolade—increased recognition at my institution, press conferences and publicity, honors that accrue not only to me but also to my family and institution, and even the rediscovery of old friends and acquaintances with whom I had lost contact.

Some years later, as chairman of Caltech's division of biology, I had the privilege to nominate Roger Sperry for the Scientist of the Year Award, which he received and most eminently deserved.

In 1969, I received the Beijerinck Virology Medal of the Royal Netherlands Academy of Sciences in Amsterdam. Beijerinck was a famous Dutch microbiologist who in the 1890s discovered the virus of tobacco mosaic, which he described as a *contagium vivum fluidum*—a "contagious, living fluid," so named because he could transmit the disease with his extracts but could see no cells or particles therein with the light microscope, nor could he eliminate the infectivity with any filters he had available.

My studies of ϕX, one of the smallest known viruses, although quite visible in the electron microscope, had clarified and extended Beijerinck's concepts at the particulate and molecular levels. The prize consisted of an engraved gold medal and two thousand dollars, which sufficed to bring me and my family to Amsterdam for the award ceremony. It was my privilege to address the Royal Netherlands Academy. As an accompaniment to receipt of the prize, I lectured at each of the five leading universities in the Netherlands—Leiden (founded in 1575), Utrecht, Amsterdam, Nymegen, and Gronigen—and was thus able to meet most of the leading Dutch microbiologists, as well as to see Holland in January.

Prizes are pleasant, but in the science of today such priority is largely an "ego trip" that gives one an illusion of leading the pack instead of merely riding the tide. In fact, nature exists and the discovery is waiting to be found; if scientist A does not find it one year, scientist B will the next. A does so first because, most often unwittingly, he or she has chosen the more accessible instance or has access to the more relevant techniques. The real *inventors* are those who create and develop new techniques—the ultracentrifuge, gel electrophoresis, the electron microscope, the scintillation counter, and so on. But most often they are not the ones to make the biological applications and thereby receive the acclaim for discovery. Yet without them, molecular biology would not exist.

The most famous prizes are the Nobel Prizes. These differ from most others in several ways. They are highly remunerative. They are awarded only in certain fields, designated by Nobel, and are awarded primarily for *consequence* of a particular discovery, that is, whether the work recognized has proven to be of seminal importance for further development in the field. Sometimes consequence can be quickly recognized, sometimes it can be delayed as long as forty years. The prize is not awarded, necessarily, for imaginative ingenuity, elegance of experimental design or execution, sustained contribution, or even (always) correctness. As a result, recipients of the prize are not always as yet members of the national academy (an oversight usually quickly corrected).

I have known almost all of the recipients in biology and medicine, and many of those in chemistry, since 1950. In my experience, the recipients of the prizes are good scientists, some very good scientists. But in nearly every instance, one could name half a dozen others of equal ability whose contributions, averaged over their professional lifetimes, would rival those of the recipients. It might be argued that the recipients should be specially rewarded precisely because they had the prescience to work on problems that had great consequence. But in many cases, this could not have been known in advance. In others, it was obvious in advance that a solution to that problem would be of great importance, and the prize winners were those who happened to choose a more favorable system to explore.

Understandably, the prizes tend to recognize those who make dramatic advances but often overlook those whose prior advances and inventions made the "breakthrough" possible. Thus, Gilbert and Sanger deservedly received the Nobel Prize for their development of means to sequence DNA, but no award was given to those who invented and perfected the technique of gel electrophoretic separation that made their work possible. Likewise, several Nobel Prizes have been awarded for research involving "recombinant DNA," but not to Boyer and Cohen who invented this technique.

Nobel Prizes probably served a different function in an earlier era, when scientists worked more as individuals and science was less connected and scientific progress more sporadic. Today, the achievements recognized by Nobel Prizes are mostly only slightly higher crests in the advancing waves of scientific progress.

However, the publicity of the Nobels creates its own aura. Even major institutions feel obliged to pay obeisance to the selections of the Swedish Nobel Prize committees. Each spring at Caltech, as biology chairman, I prepared a schedule of proposed biology faculty salaries for the next year to discuss with the provost. It was understood that any Nobelists would be at the top of the list. Later, all of the division chairpersons would gather with the provost to review all faculty salaries, to seek to achieve rough parity between the divisions, and to recognize any faculty whose contributions may have been split between divisions. In accord with institute tradition, the Nobelists in physics were always at the top of the institute list and all other salaries were pegged to theirs.

The prize doubtless serves an important public relations function for the scientists and institution concerned and for the image of science to the general public, just as Oscars do for the movies. But within their respective professions, each is viewed very differently.

16

Transition 4

At some time in my late forties I realized that I had "made it." My research was internationally acclaimed. I was the chairman of the best biology department in the world. I no longer needed to "prove myself" to myself or to others. I had achieved in my research and career what I had set out to achieve so amorphously many years earlier.

In a sense all of that accomplishment was like the uncoiling of a spring that had been formed very early in my life—the realization of values and goals imprinted as a child and focused at MIT. It was as though the program had come to the end of the tape. But where to now? With the imprinted goals achieved, what came next? Some would call this a mid-life crisis, but I had no sense of crisis but rather one of relief and decompression. Long-repressed interests outside of science reemerged.

The tone and ambience of science were changing—the result of its own success. When I entered science, it was in the popular mind an arcane but benign profession. It was peopled largely by scholars who had a driving curiosity to know how the natural world worked, and who could be nothing else but scientists. As a career, science did not promise much material reward but offered great intellectual satisfaction.

The Second World War had convincingly demonstrated the value of science to national security, health, and industry. Government funding then greatly augmented the resources available to science and with them the potential of careers in science. During the 1960s, I observed the entry of a somewhat different type of science student—bright but not driven—who chose science as an acceptable professional career occupation but who might just as well have become a lawyer, physician, developer. Between 1948 and 1988, the number of Ph.D.s awarded per year in the biosciences increased more than eightfold.

Science could now be performed on a scale commensurate to the problems at hand with a resultant acceleration of progress. And as science and its offspring, technology, became more and more important in the economy, as society was required again and again to adapt to new technological change, and as some of the impacts of the new technologies (e.g., nuclear weapons, pollution, toxic waste, even overpopulation) appeared less than desirable, so the societal ambience of science changed.

Scientific administration proliferated. The allocation of resources for science inevitably became at least partially politicized, and the societal view of science and technology became increasingly critical, whether a backlash to unsought and unwanted change, a response to perceived environmental degradation or military emphasis, or an expression of a latent anti-intellectualism. The public support for science in a society largely ignorant of its content becomes inherently volatile, resting on the public perception of its consequences. Science is still perceived as arcane, but now it is also considered potentially malignant.

These observations led me to an heretical thought: Should there—can there—be limits to inquiry? The very question incites distress and revulsion in a scientist, whose credo must be that knowledge is good and more knowledge better. But the logic of the recombinant DNA controversy forced me to consider the question as dispassionately as possible. Had it ever arisen before—among scientists, that is, not among those committed to religious or ideological dogma? Yes, it had, to nuclear physicists. Frederick Soddy, a colleague of Lord Rutherford's, had written in 1920:

Let us suppose that it became possible to extract the energy which now oozes out, so to speak, from radioactive material over a period of thousands of millions of years, in as short a time as we pleased. From a pound weight of such substance one could get about as much energy as would be obtained by burning 150 tons of coal. How splendid. Or a pound weight could be made to do the work of 150 tons of dynamite. Ah, there's the rub. . . . It is a discovery that conceivably might be made tomorrow in time for its development and perfection, for the use or destruction, let us say, of the next generation, and, which it is pretty certain, will be made by science sooner or later. Surely it will not need this actual demonstration to convince the world that it is doomed if it fools with the achievements of science as it has fooled too long in the past.

And many of the physicists who worked at Los Alamos on the atomic bomb during World War II had deep qualms. As Victor Weisskopf has written: "Many of us hoped that the number of neutrons per fission would be low enough to prevent the making of a bomb. But it wasn't."

To pose the question invites a paradox. How can we know what we would not want to know? The paradox is compounded by the tautology that the most important discoveries are those least expected.

We accept limits on modes of inquiry. We do not experiment on involuntary human subjects; we seek to minimize suffering in experimental animals (some would ban all animal research); we avoid experiments that might produce irreversible environmental effects. Implicitly, we value human dignity and our sense of responsibility for other life forms and for the planetary environment more than the knowledge that might thus be acquired. Most often, we assume or hope that the knowledge sought can be obtained by other more acceptable means.

But is there any knowledge we would not want to have, regardless of how it was obtained? Knowledge too dangerous, too destabilizing? Our species evolved and survived with the means to cope with the dangers of a world of human scale. At that scale, the species could tolerate the not insignificant level of irrationality that we know to be present in human behavior. But beneath the surface of that human-scale world lie structures and forces of great power, which we have learned to understand and manipulate to human purpose. Could there be elements of these powers that would be mortally dangerous if available to irrational minds? Does nature set traps for unwary species?

If hydrogen bombs could be readily made in a garage workshop, all of human society would be in mortal peril. Happily, it is not that easy. But would we encourage, or allow research to make such simple manufacture possible? Does biology, through recombinant DNA, offer a potential analog? And, if so, could inadvertence, let alone malevolence, unleash disaster?

I was, however, as I am today, most reluctant to concede that humanity should forego knowledge of any kind forever. I therefore thought to introduce another dimension into consideration—time. Discoveries that might be of great potential harm in one era might be innocuous in another. In a truly peaceful world, the discovery of atomic energy might never be applied to weapons. Should then more research and intellect have been devoted to the search for routes to lasting peace and less to nuclear physics?

Is there a preferable sequence for discovery, just as there is a necessary sequence for the onset of gene expression during the development of an organism? Are there more opportune times for the exploration of certain areas of knowledge? Could we discern such a program and apply it to the allocation of research resources?

These seemed to be meaningful questions, worthy of sustained exploration. These emerging concerns needed to be addressed by minds well informed and perceptive as to likely future developments. But the best minds of science were focused elsewhere. Caltech, MIT, and the national academy were largely ignoring this rising tide, except when it directly affected their specific interests.

17

Science and Society— Toward Wider Horizons

We are at a hinge point in evolution. Some two hundred thousand years ago, *Homo sapiens* emerged on this planet—out of a panoply of life— mostly mute, unself-aware, living, multiplying, dying. We emerged with this body, this frame, these hands, this mind, unaware of any past, ignorant of our inner selves, innocent of the future.

Somehow we found our way. We invented language, writing, agriculture, industry, social organization, and science and only in the last hundred years or so have we begun to learn about the physics and chemistry of our inner selves and consequently about our long ago past; about our intricate machinery and why it can go awry; about the inheritance that makes worms become worms, birds birds, and us humans.

What a wonderful time to be a biologist.

The mysteries fade in the light of knowledge, but with knowledge and understanding can come the urge to intervene, to change, to modify. But the human mind, which is so good at the analysis of what is, falters before what ought to be. This is hardly surprising. Our mind did not evolve to cope with such questions. Evolution has not confronted such issues until now.

Curiously, my conscious reflection on this profound change in the human condition all began with a request to give a Monday Night Lecture. The invitation seemed to tap a previously unconscious, unexpressed vein of interest and concern. By 1965, I had been immersed in science for nearly three decades—a student, a researcher, a teacher, I had rather singlemindedly pursued a scientific career, interacting with other scientists, teaching future scientists. Now several events were to change my path in the coming decades toward much wider and hazier horizons.

The Monday Night Lectures (now called the Watson Lectures) were a Caltech tradition. On most Monday evenings during the academic year, a Caltech faculty member would attempt to describe his or her research—methods, objectives, results—in terms understandable to an "educated" public. The lectures were one hour, followed by a question-and-answer period. Most made considerable use of visual aids. Over the years, a considerable audience of five hundred to one thousand persons had developed for these lectures, including many of the Caltech faculty. They were a successful means to reduce possible estrangement or tension between the institute and the resident community.

Having never before attempted to present my research program to a nonscientific audience, I took the "challenge" seriously. My title was "The Book of Life." I endeavored to compare the genetic information to the information in a book—like a book of recipes or a manual of flower arranging—written in an unknown language, like a Mayan codex, and I sought to explain how far we had come in deciphering this text and what it meant and what it would mean when the knowledge was more complete. I had made many diagrams for the purpose.

The night of the lecture arrived during one of the torrential downpours Pasadena can have in the winter. I anticipated minuscule attendance, but the audience was loyal and a good crowd appeared. To my pleasure and surprise the lecture was well received. Indeed, the substance was later printed by Addison-Wesley as a "separate" and sold many thousands of copies.

The lecture had another consequence. In the fall of 1966, Caltech was celebrating its seventy-fifth anniversary in Pasadena. A major symposium was planned on the subject of the future of science. Because of my Monday Night Lecture, I was asked to present a lead address on the future of molecular biology and its potential implications. I had never given this larger theme much thought. Scientists are generally immersed in their scientific questions, in daily research and the planning—invention, really—of the next experiment. The larger, more cumulative impacts of scientific knowledge are not often within their ken, and in any case seem rather speculative and remote.

But for the next six months my mind kept returning to these questions and I found I would jot down notes for this talk at all times—

early in the morning, during a car ride, at a concert intermission. It was during this time that I came to realize that the advances continually being made in genetics would before long have the most profound social consequences. I could not foresee the specific pattern of scientific advance that would lead to the potential for genetic intervention. But nature clearly had developed means for genetic modification and re-combination and we would surely, in time, and sooner rather than later, unravel these mechanisms and have this capability at our disposal.

We were at an epochal moment, not only for our society or for *Homo sapiens* but for all of life on earth. For the first time in the long course of evolution, for the first time in all time, a species was coming to un-derstand its origins and its inheritance, and with that knowledge would come the ability to alter its inheritance, to determine its own genetic destiny, as well as that of other living species. Through DNA, biology was moving beyond analysis to synthesis.

This was a transformation without precedent. The issues that would be raised far exceeded the boundaries of our historical morality born of human experience. A few novelists, Aldous Huxley for one, had jousted with the issues, but now fiction could become reality. What principles should guide us? How should decisions be made, how could agreement be reached in a fractured world society? In this light, the links between our social structure and the givens—to date—of human existence be-came evident. Our size, our life span, our numerical equality of male and female, our genetic diversity and accompanying normal range of talents and traits, all are imbedded in our social order. And all now are potentially mutable.

The potential for intervention in human inheritance raises funda-mental questions of the meaning of "equality of opportunity," of re-sponsibility (liability?) when genetic chance is replaced by design, and of human dignity when an individual realizes that his or her traits have been programmed.

The prospects both exhilarated and sobered. I was exhilarated by this triumph of human intellect yet sobered by the need to use wisely the powers thus unleashed. I have been less innocent ever since.

My lecture was entitled "The End of the Beginning" to connote my perception of this turning point in evolution.

Because this message was ultimately minatory, I was apprehensive that it might not be well received by a Caltech audience, but the au-dience seemed taken by the revelation of the extraordinary significance, new to them, of the recent advances in genetics, and the talk was very well received. Indeed, Thomas Watson, Jr., then president of IBM and a Caltech trustee, invited me to Bermuda to present the same talk to his IBM administrative retreat. In the succeeding years, I was invited to present numerous talks expanding on this general theme. Some of the titles were a bit pretentious—"Darkly Wise and Rudely Great" (from Alexander Pope); "All Men Are Created Equal?"; "The Brain of Pooh: An Essay on the Limits of Mind"; "An Inquiry into Inquiry"; "Humanism and Science"; "The Galilean Imperative"; "Prospects for Future Scientific Developments: Ambush or Opportunity."

One of my lectures, "Science and the Quest for Human Values," was at the Pittsburgh Theological Seminary in connection with the 175th anniversary of their founding. While there, I visited their library and was astonished to find a collection of some eight hundred regularly published journals from all over the world. There were only two I had ever seen. Here was a whole universe of scholarship of which I was completely unaware. At Pittsburgh, I began:

In time it will probably be seen as inevitable that science, which set out simply to explore the universe objectively—without the constraints of, indeed orthogonal to, the concerns of value—should have come to test in the harshest way the fabric of our values. Today it takes little vision to see that science is ready to pose to man wholly unprecedented questions of the most fundamental character which will of necessity require a reformulation and a deeper understanding of our basic moral principles.

For us in science, it is frankly still surprising to have come from a new direction upon the oldest of questions. Perhaps, as we reflect, this consequence will tell us something about the geometry of fate and the matrix of the human mind.

We are already confronted with grave dilemmas arising from the only partial triumphs of science, which splinter our older values and often expose their expedience and inconsistency. The Malthusian tide of population mocks our belief in the value of individual man; the flood of factual knowledge overwhelms our faith in the immanent value of truth and thrusts us in the feckless role of the sorcerer's apprentice; the disintegration of death undermines our ancient views of the beneficent role of the healer and the worth of human life. Graver questions yet lie ahead as the biological sciences prepare to change the boundary conditions of man.

These talks brought me into contact with a wide variety of people and I began to appreciate how extremely varied were the reactions to my message, how each sought to incorporate it into his or her frame of reference, with resulting fear, awe, distaste, greed, eagerness, resignation, antipathy, consternation, or dismissal. Among scientists, the reaction was most often one of exclusion. "Our mission is to acquire knowledge. How it is used is not my concern." Or one of temporizing: "These issues may arise at some time in the future but they need not concern us now." Or of apprehension that raising such questions now might jeopardize current research funding. These responses reflect in part the narrow specialization of much scientific education and in part the intense dedication required of an active life in science.

Continued reflection raised broader questions in my mind as to the largely self-directed patterns of scientific exploration, mostly government funded. Scientists, understandably, want to pursue their own research inclinations. And surely at the research project level, who should know better what paths are the most promising? Out of this perception has arisen the "peer review" process in which proposals for research funding are evaluated by panels of scientists expert in each field, ranked in order of promise, and funded as far as available resources permit.

But does the process assure on a more macroscopic level that funds are directed toward research on problems most pressing or crucial to society in general? It seemed to me that this result was not obvious— that, for instance, in a world confronting large and unmanageable increases in population, it might be desirable to place more resources into means of fertility control and less into studies of the aging process. Or more funds might be spent on studies of drug addiction, and less on novel devices for military use. Such issues of science policy should not and could not be resolved solely by scientists, but a forum was needed within science for their discussion. None was, or is, available. I presented many of these thoughts in an article entitled "The Presumptions of Science, " published in *Daedalus.* It did not win me many friends in the scientific community.

As a result of these contacts and cogitations, when the recombinant DNA issue arose in the mid-1970s, I came to it from an atypical perspective.

The development of the recombinant DNA technology provided us with the ability to move genes about from one organism to another almost at will. The universality of the genetic code implied that a gene, a tract of DNA, could engender synthesis of the same protein wherever it was placed, provided of course that it was activated in the host cell. This concept provided the potential to use microorganisms as biochemical factories to produce scarce proteins of pharmaceutical value in medicine, such as human insulin, blood-clotting factors, growth hormones, interferons, and so on.

For molecular biologists, this development provided the means to generate for study large quantities of any gene of interest, known or unknown. And, of course, it provided the means, by introduction of specific genes, for intervention in the developmental and biochemical processes of plants and animals to alter these for human purposes. In short, genetic engineering was here.

Was it possible, however, that these experiments might, by design or inadvertence, produce an organism of undesirable properties, one that could be a danger to man or to the ecosystem? In the long course of evolution, genetic exchange among unrelated species is a great rarity. The evolutionary pattern is a branching tree, not an interlaced network. We would be creating novel genetic combinations never before seen on earth.

Much of this work would be performed, initially at least, on the bacterium *E. coli* simply because its genetics was so well understood after thirty years of intensive exploration. But our laboratory *E. coli* strains had originally been derived from strains found living in human intestines. While undoubtedly attenuated, after decades of laboratory existence, might they still retain the ability to colonize the human gut, now bringing with them all manner of exotic gene combinations? A popular form of experiment at the time was the "shotgun" experiment in which quite random fragments of DNA from higher organisms were inserted into *E. coli*. The genes borne by such fragments were completely unknown. Might some, by accident, be toxic? And, of course, the potential for deliberate insertion of toxic genes, for the purposes of biological warfare or terrorism, was evident.

I was further perturbed by another class of experiments then underway. At this time, tumor viruses from various animal species—birds, cats, baboons—were being mutated, recombined, and so on. At that time, no virus causing tumors in humans had been discovered and it seemed evident that viruses were not a major cause of cancer in man. But viruses were a major cause of cancer in other species and we might "inadvertently" in our experiments generate a human tumor-inducing virus. I was also concerned by the fact that the new forms—viruses, cells, organisms—were living, self-reproducing creatures that, should they be inimical and should they escape from the laboratory to find a suitable ecological niche, could multiply indefinitely. They could not be recalled, nor would it suffice to halt their manufacture. The generation of new living organisms was qualitatively different than the generation of new inanimate objects or chemicals.

Some of these concerns had been raised within the biological community. A "moratorium" had been suggested for recombinant DNA experiments and a conference to discuss these issues and potential hazards was convened at Asilomar in February 1975. I attended this extraordinary, unprecedented, and unrepeated event. About 120 of the leading molecular biologists attended. The sessions were divided between those in which the most recent scientific developments obtained with the new techniques were discussed with great enthusiasm (what moratorium?), and those in which the conceivable hazards were described and means considered to mitigate them.

Molecular biologists, including myself, trained mostly in biochemistry and biophysics rather than pathology, had largely treated viruses and bacteria as we would any other chemical, taking enough precautions of sterility to accomplish our own experiments but otherwise exposing products to the air, pouring used cultures into the sewers, and so on. Without much data, a wide range of opinions was expressed with regard to the conceivable hazard of various genetic introductions. A discussion by a panel of lawyers informed the audience as to the liabilities to which they might conceivably be subject.

Means for the containment of recombinant organisms within the laboratory were discussed. Physical containment in sterile rooms with air locks was feasible but costly and troublesome. The concept of biological containment, the use of strains of *E. coli* deliberately weakened so as to reduce their likelihood of survival outside of carefully controlled laboratory conditions, gained favor. Such strains were not available but could plausibly be made within a few months. (This was an erroneous assumption—it proved to be much more difficult to create and work with such strains.)

Clearly, there was far less interest in the discussion of hazards than of new science. Any means to cope with such potential hazards would inevitably hinder research. Opinions ranged from fear of possible epidemic to cavalier statements that if a pathogen should result and a few persons die, it would only cost a few million dollars. In the end, a compromise was reached in which various conceivable experiments would be graded as to their potential hazard. Combinations of physical and biological containment would be proposed, commensurate to the perceived hazard of each experiment. The restrictions would range from minimal on the most innocuous experiments to bans on the most dangerous. The details would be worked out by an NIH committee.

It was all plausible; however, since no one knew the absolute level of hazard, no one was sure just where to "float" this entire matrix of regulation. To nobody's surprise, it was in fact adjusted so that the experiments that most scientists wanted to do at once could be done with minimal precautions, thus buying time to provide more elaborate and safer facilities for experiments to be done in the future. Science could go forward in good conscience.

In the year or two after Asilomar, I became increasingly uneasy with the safety of the regulatory mechanisms developed. They seemed to me to be token. If the problem was to be taken seriously, quite other approaches would be desirable. Instead of using an organism such as *E. coli* derived initially from human waste and capable of exchanging genes with other organisms known to reside in the human intestine, a host organism such as a thermophilic bacterium—which could only survive at very high and uncommon temperatures—might be used. Sophisticated, high-containment facilities might be established at ten or twelve centers around the country at which the more hazardous experiments could be performed.

Others shared this view, but few were willing to speak out against the biological establishment. Because my position was secure, I felt an obligation to do so. On almost every occasion that I did, others, often junior faculty, would let me know that they agreed with my position but dared not to speak out. They feared reprisals from more senior and powerful colleagues who controlled funds and advancements.

The controversy over DNA regulations became more intense and, as often regrettably happens, was joined in by less than rational and ill-informed nonscientists bent on ideological or political purposes—the Jeremy Rifkins and the religious zealots. This was my first experience with unwanted fellow travelers. It is not always pleasant to be associated, in common cause, with others whose motives one considers suspect or absurd.

Over time, two developments quite changed our perception of the problem. The *E. coli* strains in laboratory use proved to be considerably more attenuated, less able to compete with indigenous strains outside the laboratory, than might have been supposed. And even more important, the processing of genetic information, of messenger RNA, in higher organisms unexpectedly proved to be much more complex than in bacteria. As a result of the discovery of introns and splicing mechanisms, it became clear that most genes, simply extracted from a higher organism and placed without modification in a bacterium, could not give rise to a functional protein. While complicating genetic engineering, this discovery greatly reduced concerns about hazards from "shot-gun" experiments.

I continue to believe that, with the knowledge available at the time, my concerns over the potential medical and ecological hazards associated with the introduction of the recombinant DNA technologies were fully justified. We lucked out again. But my opposition to the biological establishment in this matter was personally costly. As one of the few opponents who could not be readily dismissed, I became a target for enmity. I think it no accident that since that period I have not been asked to serve on any NAS or NIH committee, although I had previously done so regularly and despite the fact that I have specifically responded to calls by the president of the national academy for its members to become more personally involved. Such is the price of offense to vested interests.

Persistent dissent is treated harshly within the scientific community. Committed to a single truth, scientists tend to dismiss and exclude those who willfully do not "see the light." Empirically based, science is a self-correcting enterprise. There have been few instances in which long-held, seemingly established concepts have been disproved. Thus, the persistent skeptic is viewed as at least eccentric, if not fanatic, and harboring hidden motives.

Throughout this period of engagement with questions of values and issues of controversy, I became increasingly concerned with the perceived failures of our educational system. The bulk of our citizenry is technologically illiterate. We are creating an increasingly technological society in a democracy with a public largely unprepared even to understand the new and complex issues emerging from our enlarged powers. And conversely, the small group of technologically informed citizenry are largely unconcerned with the social implications of their activities. Their education has not prepared them to cope with the vexing questions of social and cultural values. Indeed, many have no doubt deliberately pursued a career in science with its clarity and objectivity so as to escape the complex, ambiguous, sometimes unresolvable questions of public policy.

Scientists are not innately more rigorous or logical, but they face daily the task of comprehending an objective reality. Bias and cant, aesthetic preference and moral conviction are but snares. The natural world is as it is, not as we might wish it or would construct it. Constrained thus to objectivity as a way of life, scientists can only view with wonder and alarm, even scorn, the self-serving passions that so often pass for argument in most social and political arenas—the hyperbole that contorts logic, the myopia that accompanies self-righteous claims of superiority. The limits of evidence are far transcended; the possibility of error seems never admitted.

Scientists, on the whole, wish simply to pursue their investigations of nature—to explore, to discover, to satisfy their curiosity. Nature can be obscure, but nature does not deceive. Yet their investigations are increasingly costly. Because funding is public, the scientifically ignorant public must continually be persuaded that it should, in its best interest, support scientific programs. Thus, the entire enterprise rests on a public confidence that in turn rests more on faith than knowledge. At the policy level, then, the scientific establishment is deeply insecure. It is little wonder that its leadership feels obliged to present a united front and to resist suggestion that any aspects of the scientific endeavor might not be in this society's ultimate best interest.

This state of affairs bears the seeds of future disaster. An ignorant public and a detached yet wary scientific elite could easily be separated by determined political or ideological forces. We have now seen this happen in such diverse areas as nuclear power and animal rights.

By the mid-1970s, it seemed to me that it might be just as important for the future of science for me to devote my efforts to educational reform as to further advance scientific knowledge—to the design of educational patterns producing a more technically literate citizenry and, particularly, a more socially aware scientific cadre.

But where to do this? Caltech, trapped in its own scientific success, dedicated to the most advanced scientific research, was not a promising site for curricular innovation. Nor could the institute be substantially broadened. In the early 1970s, the American Academy of Arts and Sciences was looking for a new academic site in which to establish a center for advanced study in the humanities. I had suggested Caltech. The academy leadership was truly intrigued by this suggestion for possible liaison between a humanities think tank and a small preeminent scientific institution. Conceivably, "two cultures" could produce a dramatically successful synergism.

But the proposal, while not rejected by the Caltech administration, was not pursued with enthusiasm, and the center went to North Carolina instead. A pity!

18

The Caltech Years—
The Chairman

"I think you mean 'million,' Harold." Harold Brown and I, as the new president of Caltech and its chairman of biology respectively, were going over the biology division's budget for the next year. Until very recently, Brown had been secretary of the air force with a budget of some thirty billion dollars per year. The Caltech budget at that time was about thirty million dollars per year. Since the budget tables provided only the first three or four numbers, omitting the trailing zeroes, his numerical confusion was understandable. After the second correction, I stopped. The zeroes would soon enough correct themselves.

To become president of Caltech or chairman of its division of biology is to join a distinguished roster. Thomas Hunt Morgan became the first chairman of the division of biology at Caltech in 1928. Morgan, who received the Nobel Prize in 1933 for his brilliant studies of inheritance in the fruit fly, drosophila, was the most eminent geneticist of his time. He and his research group had been at Columbia University for more than two decades. That Robert Millikan, Caltech's president, was able to persuade Morgan at the age of sixty-two to uproot from Columbia to a small, little-known institute on the then-remote West Coast surely bespeaks a most persuasive man.

Millikan had raised a million dollars from a neighbor, the industrialist William Kerckhoff, to build the Kerckhoff Biology Laboratory and to endow the new division of biology. Caltech had been created in 1920 out of the existing Throop Institute (a practical trade school for training machinists and nurses) by three distinguished physical scientists, the astrophysicist George Ellery Hale, the chemist Amos A. Noyes, and the physicist Robert A. Millikan. (Both Hale and Noyes had been associated with MIT.) Biology, which was introduced arbitrarily by Millikan, was a distinct departure from the earlier concentration in physical science and mathematics and is reported to have produced some faculty grumbling: "Biology! Will theology be next?"

Millikan and Noyes foresaw that biological processes would be increasingly amenable to chemical and physical investigation and it was this vision, and the potential for its realization at Caltech, that was persuasive to Morgan. It is curious, however, that the first Kerckhoff biology building did not have a single fume hood for chemical research.

Millikan was peculiarly astute in his selection of a geneticist to head the new division. Genetics was destined to become the central theme of biological research in the twentieth century (and likely beyond), but this was not obvious in the 1920s when experimental embryology and physiological biochemistry were in their heyday. Morgan brought his drosophila group with him, including Henry Sturtevant, Calvin Bridges, and Sterling Emerson. California was then considered so remote that these faculty were assured they could return every other summer to the marine laboratory at Woods Hole, expense paid, to maintain their scientific contacts.

After Morgan's death in 1943, the chairmanship was held by temporary appointments until Lee DuBridge brought George Beadle, who received the Nobel Prize in 1957 for his studies of biochemically deficient mutants in the mold neurospora, from Stanford. Under Beadle, the division began slowly to grow and diversify. Notable arrivals included Delbrück in 1948, Ray Owen, a geneticist interested in immunology, in 1946, and the neurobiologist Roger Sperry in 1954. Beadle left in 1960 to become president of the University of Chicago and was succeeded by Ray Owen. When Owen decided to step down in 1968, I was asked by DuBridge to succeed him.

In 1968, after twenty-two years as president of the institute, Lee DuBridge was approaching retirement. He had been an outstanding administrator. In his early years, he revitalized the institute. His continuing good judgment in science had retained the confidence of a somewhat arrogant faculty; his warm manner, his ardent conviction of the value of science, and his exceptional ability to present the aims of science in popular terms won broad local and national support for the institute. While accepting of specific tasks, he, unlike many of his confreres, had eschewed long-term commitments to the Washington scene, believing that Caltech required his primary attention and energies.

In his later years, the success of his leadership in turn led to a gradual growth of institute activities, which outgrew his capacity for direct supervision. His personal style of administration did not permit him to develop an appropriate administrative substructure. Thus, toward the end of his tenure, the institute was in some disarray as regards finance and administration.

As a private institution, Caltech is formally governed by a self-perpetuating board of trustees. Their most important function is the periodic selection of a president. Having done this, a wise board—and Caltech has a wise board—leaves the academic administration of the campus to the person they have chosen. They continue to keep themselves informed, they maintain a close watch on the financial status of the institution, and they serve as a source of and conduit to the major gifts essential to the survival of all private schools.

In looking for a successor to DuBridge, the trustees were evidently seeking an individual with scientific training who also had the extensive managerial experience then needed to realign the institute's finances and administration. Good contacts in Washington, the primary source of research funds, were also desirable. The two leading candidates were James Fletcher, who had been director of NASA, and Harold Brown, former director of Livermore National Laboratory and the current secretary of the air force. After the election of Richard Nixon in 1968 and the onset of a new Republican administration, options changed quickly. DuBridge was asked by Nixon to become his science advisor, and Harold Brown obviously needed a new position and became instantly available.

His appointment as science advisor seemed a most fitting cap to DuBridge's distinguished career. In fact, it developed otherwise. Unable to cope with the hardball, partisan tactics of the Haldeman-Ehrlichman shield around Nixon, DuBridge retired after two years.

Because of his military connections, Harold Brown's candidacy for president was regarded with some suspicion by Caltech faculty. An intensive series of meetings by Brown with various faculty groups allayed most doubts. With the urgency generated by the immediacy of DuBridge's departure, the trustees quickly selected Harold Brown to be the third Caltech president.

He proved to be an excellent administrator with rather little educa-

tional vision. He was content to let the faculty develop their own concepts of desirable research and educational directions. At an institution
of the calibre of Caltech, this policy works for a time. Even at Caltech,
however, faculty tend to reproduce themselves and their programs and
a guiding, innovating hand is occasionally needed. Unquestionably brilliant, Brown lacked the personal warmth of DuBridge and never
achieved similar personal rapport with faculty or supporters.

As president, Brown chaired the Institute Administrative Council
(IAC). Because Caltech is small, the chairmen of each of the six disciplinary divisions of necessity also play a major role in the relatively small
central administration. All major policy questions are discussed in the
IAC, composed of the president, the provost, the six chairmen, and the
vice-presidents for finance and for development. Thus, a division chairman became well acquainted with issues of institutional policy as well
as with the concerns of his own division. Chairmen differed widely in
their abilities to adopt an institutional point of view as distinct from the
parochial interests of their own constituencies.

Under Lee DuBridge's more autocratic style, the IAC meetings were
not so much forums for discussion as conventions to confirm decisions
previously made in his office. Harold Brown had developed a different
style while secretary of the air force in dealing with military command-
ers. Under his leadership, meetings of the IAC were much more vigorous, although the final decisions were clearly his to make. Men of
eminence in their respective fields, these were no "shrinking violets,"
and discussion was intense, albeit respectful. Jack Roberts in chemistry
was especially vociferous in defense of his views. Bob Huttenback, later
to become chancellor at the University of California, Santa Barbara, was
an effective advocate for an expanded role for the social sciences at
Caltech. The meetings were frequently enlivened by differences between Robert Christy, the provost, and William Corcoran, the vice-
president for development. Christy, a physicist, was direct and tended
to see issues in black and white. Corcoran, a chemical engineer by profession, spoke in a very prolix manner, spiraling about his subject in ever
narrower circles until he was finally ready to make his point. His perceptions were usually complex, even baroque. The two distinct visions
and styles often clashed, producing sparks.

The IAC was concerned with academic, fiscal, and administrative
issues. The problems confronting the institute in that period were significant but not critical. All tenure decisions in all disciplines were discussed there. The academic-research programs of the divisions were

reviewed. Those in biology, chemistry, earth sciences, and astronomy were considered eminent and in good hands, with sufficient opportunities for renewal with young appointments. Physics, especially theoretical physics with both Richard Feynman and Murray Gell-Mann, was outstanding; the lack of significant activity in solid-state physics did not seem troubling.

The mathematics program, however, was not considered to be of a stature commensurate with other facets of the institute and several efforts were made, unsuccessfully, to recruit leading mathematicians. An "ideological" split between the "pure" and "applied" mathematicians further complicated this area.

Engineering at the institute was a more severe problem. Student interest in engineering had ebbed to a low of 25 percent of the student body. While there were several outstanding faculty in various fields, it was widely felt that the institute had "missed the boat" in the development of computers and the associated technologies, although to some extent this was compensated by the activities at the Jet Propulsion Laboratory. The engineering faculty was aging, but there were few retirements and therefore limited opportunities for faculty renewal. Successive chairmen wrestled with this problem with limited success in this period.

Modest efforts were also made to broaden the institute's offerings in the social sciences, particularly in the more quantitative areas of economics and organization theory.

The IAC considered other, more immediate, often vexing issues of overall institute policy, including the provision, in a financially feasible manner, of adequate computing capability for the campus. Sharing facilities with JPL proved to be a salutary solution, though campus-JPL relations in general required continuing attention. The issue of classified research at JPL, undertaken at specific government request, was troubling especially during the Vietnam period of unrest. The institute received a modest but not insignificant "management fee" for its operation of JPL. The insistent temptation to build this into the institute budget had to be resisted so as not to make the institute beholden to continuation of this relationship, should that become undesirable. Periodic changes in "overhead" provisions on government grants and contracts produced corresponding fluctuations in the institute budgets and funding for graduate student support, which had to be met with one or another stratagem.

The institute's astronomy program, together with the astronomy di-

vision of the Carnegie Institute of Washington, operated the Palomar
Observatory (which belonged to Caltech) and the Mt. Wilson Observatory (which belonged to Carnegie) jointly as a consortium. This arrangement worked for some years with a manageable degree of friction.
The usefulness of Mt. Wilson, however, was increasingly compromised
by light pollution from Los Angeles. The Carnegie Institute determined
to shut down Mt. Wilson and to build a new observatory in Chile. It
desired Caltech's financial participation in this project; however, at that
time Caltech did not see such investment as a high priority. The consortium subsequently dissolved and Carnegie went ahead on its own
with the Southern Hemisphere telescope.

For much of its history, Caltech had been an all-male institution.
Women graduate students were first admitted in 1954 and women undergraduates in 1970. The proportion of undergraduate women quickly
rose to 15 to 18 percent. At this level, however, their limited numbers
created problems of mutual support and social ambience. The high entrance requirements and the restricted range of curricular options limit
the potential pool of women applicants. In recent years, determined
efforts have raised the proportion to over 30 percent, a more viable
ratio.

This IAC experience with institutional problems in higher education
was an invaluable, if only partial, preparation for what was to come.

As chairman, I was expected to attend the annual fall trustees' meeting, which was held off campus, usually at a ranch near Palm Springs.
The Caltech trustees are a remarkable group, involving many distinguished citizens, including the chief executives of several of the most
technologically innovative companies. While the trustees are concentrated in California, a deliberate effort has been made to include trustees
from all parts of the U.S. and from a wide range of activities and industries. They included such civic and industrial leaders as Judge Shirley
Hufstedler, later secretary of education; Robert McNamara, former secretary of defense; Thomas Watson, Jr., president of IBM; Arnold Beckman, president of Beckman Instruments; Simon Ramo of Thompson-
Ramo-Woolridge; and so on. In addition to providing a broad range
of expertise, this composition also gave access to a wide field of potential
supporters.

As leading executives, the trustees are very often strong personalities.
Brown, having coped with top military brass, was able by sheer intellectual prowess to manage and direct trustee meetings toward his desired ends. I watched with admiration as he verbally sparred with and

subdued a rambunctious Fred Hartley, the irascible head of Union Oil, or corralled an outspoken Howard Keck, the self-made, blunt-speaking oil magnate.

These kinds of issues and the accompanying milieu were distinct from those to which I had become accustomed in research and teaching. Why did I want to become chairman? The responsibilities, while not excessive, would cut into the time available for research, editorial, and other professional roles. Also, I was assuming a major implicit responsibility for the future welfare of the division and my colleagues. Partly, I felt honored that my colleagues, and the president, would entrust me with this role, to follow such distinguished predecessors. Partly, I was curious about the practice of university administration. Partly, I realized that a number of the biology faculty would be retiring in the next decade. The choice of their successors would be critical to the quality and direction of the division. I felt ready and qualified to provide this leadership. And, of course, if I did not accept this post, who would be the next chairman, and would I be satisfied under his leadership?

As chairman, I soon found that administration required a quite different pattern of cognitive skills. Experimental research requires an intense, almost single-minded focus on the question at hand. One lives with the problem, seeks for alternative explanations of the data, searches for a coherent generalization that can be tested. Administration, however, requires instant shifts of attention and quick shuttles from one memory bank to another as different issues are brought for one's consideration. One's thought is more reactive, less self-generated. Indeed, it is rare to find time to think extensively about any one issue. At first, this transition was quite difficult for me. Once it was made, however, going back to the previous pattern was even more difficult.

As a professor, I had shared the usual faculty perception of university administration: an unfortunate evil, seemingly necessary to provide an interface with the outside world, often dedicated more to its own interests than to those of the faculty. A good administration was one that intervened minimally with the wisdom of the faculty. I soon learned to recognize the myopia of that view. The administration does buffer and safeguard the faculty from the external world, but just as much, it protects the faculty from its own rampant self-centeredness. By maintaining civil and orderly processes, it forestalls entropy and preserves the faculty from drifting into anarchy. Even at Caltech, difficult decisions must be made concerning the allocation of resources, the distribution of emphases, and the boundaries of propriety. Wise leadership must seek to discover the directions of the future amidst the importunities of today.

At its external face, the administration must raise funds, comply with government regulations, cope with lawsuits, and seek to maintain good public and community relations within a society that has little real understanding of the institution's activities or fundamental purposes.

As chairman, I participated in biennial joint meetings of the Caltech and MIT administrations. At these meetings, each discussed the issues it was facing, many of which were, of course, common, such as interactions with the federal government, overhead charges, relations with industry, policies regarding entrepreneurial faculty, patent policies, curricular issues, recruitment of students (especially women and minorities), and fiscal policies, and the often distinctive ways in which they sought to address these. It was most interesting to me, as a former student, to see this other, administrative side of MIT. Much larger than Caltech, more hierarchical, more bureaucratized, more involved with government and industry, the formal MIT style and the personal Caltech style were quite different, although both shared the same problem-solving approach and often, by different routes, came to similar conclusions.

Because of its greater diversity, its closer ties with the military, and its sheer size (which made it easier to gather a critical mass of activists), MIT was not spared the student turmoil of the 1960s. The administrators' accounts of their difficulties, and the efforts needed to cope with the issues, made us grateful for our more benign circumstances.

To be chairman of the biology division at Caltech is to occupy a position of prestige and wide notice. One is quickly asked to perform a wide variety of national services. In 1970, I was elected to the Council of the National Academy of Sciences for a three-year term. In the same year, I was elected president of the Biophysical Society. In 1971, I was appointed to the advisory committee to the director of the National Institutes of Health for a three-year term. In 1972, I became a member of the scientific advisory board to the Jane Coffin Childs Scholarship Fund for a four-year term (which was later renewed). I served for four years on the advisory board to the Scripps Institute and for four years on the scientific advisory board of the Merck Pharmaceutical Corporation.

As almost all of these activities were nonpaid, in effect Caltech subsidized these contributions of my time. Requests for such outside service can become excessive. I had to decline some, such as service as an alumni member of the MIT Corporation, because I was already making two trips per month to the East Coast. Obviously, my research, my teaching, the biology division, and Caltech itself required some attention.

It is important for the chairman to be alert to the advances and developments in biology broadly so that he can provide leadership and guidance to the division and the central administration when it becomes necessary to replace retiring faculty or when the opportunity arises to add new faculty positions. To accomplish this, I read widely; attended diversified meetings, such as those of the American Association for the Advancement of Science, or specialized meetings, such as those of the New York Academy of Sciences, that I otherwise would likely have passed over; and, when visiting other universities to lecture, made it a point to learn broadly about their programs and plans. My activities at the national academy (especially the editorship of the *Proceedings*) and at NIH helped to inform me of the latest progress in varied fields.

In the early 1970s, it was becoming clear that the great discoveries of the 1950s and 1960s in molecular biology, which were largely achieved through studies of simple microorganisms, could soon be applied to the study of the much more complex cells and processes of higher organisms. The development of recombinant DNA technology and cloning methods in the early 1970s solved the long-standing difficulty of obtaining an adequate amount for analysis of any one gene of a higher organism. Newer methods of light microscopy and major refinements of the methods of electron microscopy provided the potential for new insights into cellular substructures.

Over the years, I encouraged the division to bring in new faculty who would exploit these opportunities. We added Elias Lazarides, an expert in the area of dynamic cellular ultrastructure, a field that has steadily grown in importance in unlocking the mysteries of cell shape and movement and the varying locations of cellular organelles. The complex field of immunology also seemed ripe for attack by molecular methods. We brought back Leroy Hood, a former Caltech graduate destined to make major advances both in immunology and in the associated area of biological instrumentation.

Giuseppe Attardi set out to unravel the functions of the mitochondria, small bodies present in every higher cell that, curiously, carry their own small pieces of genetic material. He succeeded in completely mapping and deciphering all of the genes of this organelle.

As refinement of technique made it possible for electron microscopy to approach its inherent potential, it became a tool of increasing importance in many areas of biology. We were fortunate to bring on board Jean-Paul Revel, who established a research program and up-to-date laboratory in this field.

Developmental biology had a long tradition at Caltech, beginning of course with Thomas Hunt Morgan. After the sudden and untimely death of Albert Tyler, a student of Morgan's, we brought in Eric Davidson, who continued to exploit the classical system of sea urchin egg development. With Eric came Roy Britten, formerly of the Carnegie Institute, and together they began to analyze developmental processes at the genetic and transcriptional levels.

Subsequently, understanding of many of the genes involved in development and their roles in its early stages has advanced rapidly. Progress has been built in large part on the genetic analysis of development in drosophila, which had been painstakingly worked out at Caltech over many years by Ed Lewis. When the molecular techniques finally became available, Lewis could provide a wealth of genetic information together with the essential mutants and specialized breeding stocks.

My principal accomplishment as chairman was the establishment of a significant program in neurobiology. Caltech had had for many years a low-key program in neurophysiology with Professors Wiersma and Van Harreveld, and it had a singular program in psychobiology with Roger Sperry. I was convinced that neurobiology would be the next great frontier of biology. Armed with the techniques of molecular biology on the one hand and with the techniques of neural system analysis derivable from the field of computer design on the other, great progress seemed possible.

Dramatic progress in one field often derives from the introduction of new concepts developed in another. The availability of heuristic conceptual models can be a key requirement. Thus, understanding of the mammalian circulatory system relied on the prior knowledge of pumps and pipes. Similarly, development of computers and computer programs that could simulate at least some mental processes has provided useful conceptual models for events in the central nervous system. But the brain is not simply an intricate computer. An elaborate chemistry, genetically programmed, is required for the formation of its complex circuits and, as well, for its continuing function. In the past few decades, a wide and growing variety of neurotransmitters, modulators, and associated receptors has been discovered, all required for the effective and specific transmission of impulses between neurons.

We needed first a leader for the program. Superficially, Roger Sperry would have seemed the obvious candidate; his experiments on the specific programming of neural connections and on the distinctive functions of the two hemispheres of the human brain were extraordinarily brilliant in design and execution. Indeed, the latter, by demonstrating the existence of two separate consciousnesses in the separated hemispheres, provided one of the few true *experiments* on the nature of consciousness. But Roger had, unfortunately, the wrong personality for this role. A loner in research, his opinion of virtually all other neuroscientists was dim at best. His scale of approval ranged from minus infinity to zero. A "neutral" estimate was in fact high praise from Roger.

Of course, much is known about the outcome of brain function through simple observation, introspection, and psychological research. What is sorely needed is understanding of the mechanisms at the chemical, cellular, and systems levels underlying these observations. For this reason, I sought to find a leader who could combine a firm grasp of neurobiology with a deep knowledge of the psychological phenomena to be explained. I found him in the person of James Olds.

Olds was at Michigan and was famed for his discovery of "pleasure centers" in rats, regions of the brain for which electrical stimulation was reinforcing. Rats, given the opportunity to stimulate their own brains in these regions, were swiftly addicted to the stimulus to the degree that they would cease to eat or drink so as to continue the stimulation. The neurochemistry and neurophysiology of these regions and their connections to behavioral patterns were clearly of great interest. Olds was also interested in the processing of sensory information and in particular the conditioned filtering of incoming data by centers in sensory pathways to permit the selection of those inputs previously determined to be significant or desirable.

To establish the program, we next needed a new laboratory building. Happily, Arnold Beckman, then chairman of the Caltech board of trustees, quickly perceived the potential of this new frontier for biology and agreed to underwrite the cost personally. Caltech has repeatedly been most fortunate in the foresight and generosity of Arnold and Mabel Beckman.

With the prospect of a new building and funds for a broad new program, I was able to entice Jim Olds to come to Caltech. He quickly built up a strong group of young faculty members interested in varied aspects of neurobiology. Jack Pettigrew was interested in the role of early experience in the establishment of neuronal connections in the cat. John Allman, building on the work of David Hubel and Torsten

Wiesel, demonstrated the existence in the primate brain of multiple representations of the visual field, each specialized for a particular mode of analysis. David Van Essen explored in particular the pathways and domains involved in color perception. Jim Hudspeth studied the exquisitely sensitive manner of sound transduction into neural impulse in the cochlea. Mark Konishi sought to analyze the combined roles of inheritance and early experience in the development of bird song in various species. And Seymour Benzer elegantly undertook to identify and characterize mutants affecting neuronal function in drosophila and thus provide probes for more detailed analysis of central nervous system organization in that organism.

Tragically, Jim Olds died in a drowning accident a few years later. While the program suffered from this loss, it has recovered well and continued to develop along the directions originally foreseen.

The new Beckman Laboratory of Behavioral Biology looked across a grassy mall to Baxter Hall, the locale of the division of humanities and social sciences. The humanities and social sciences had long been peripheral to Caltech, considered a necessary part of a rounded (or at least elliptical) education, but not intellectually linked to the institute's primary thrust. I thought that such a connection might now be made between behavioral biology and the humanities and social sciences through the introduction of a program in cognitive psychology. I felt this link would strengthen both sides of the mall and would provide a new coherence to the entire institute curriculum.

I broached this concept to Harold Brown and to the entire Institute Administrative Council. The chairmen of all of the science and the engineering divisions were strongly opposed. They clearly saw such a new division as a competitor for institute resources and one that would provide scant benefit to their programs. Harold Brown took a neutral stance. The power of inertia became startlingly clear. Each of the extant divisions had a strong spokesman, while a new, unborn division had none. The necessity for strong, visionary leadership also became very clear.

As chairman, I had my first experience with what I found to be the most frustrating and unpleasant of my administrative duties—coping with personnel problems. The excellent long-time executive assistant to the chairman of biology had retired a few months prior to my appointment and Ray Owen had hired a replacement. Owen warned me that he was not sure how the new man would work out. He didn't. For several months, I tried to work with him, to instruct him as to what was needed and how it should be done, but to no avail, and in the end I had to bluntly fire him.

I also had my first, but not my last, experience in coping with alcoholism. A good friend and fine scientist on the biology faculty fell victim to this addiction. Therapists and clinics were to no avail and in the end he had to be coerced into resigning, for he was simply unable to perform his duties. This was a tragedy for himself, his family, and his friends.

But the most unpleasant encounters concerned "negative tenure" decisions, when the biology faculty had concluded that a young faculty member's performance did not merit the award of a lifetime appointment of tenure. These junior faculty had typically been with the division for six years and had naturally been treated as colleagues. As chairman, I met regularly with individual young faculty members to review their performances, to ensure that they were receiving adequate resources, and to give them counsel. In each of the negative tenure decisions, I had seen the warning signs well in advance and attempted to provide constructive advice. Such advice was rarely taken and the decision, which I conveyed, invariably came as a bitter shock. The angry, disappointed candidates were given another year's appointment during which they could seek—and, at that time, always found—another position, but it was always a year of tension. As chairman, I could persuade myself that the decision was best both for the institute and for the individual, who could find a more appropriate position elsewhere—but that intellectual theorem did not much dilute my emotional distress.

By the mid-1970s, the research with ϕX was drawing to a natural end. The essential features of the viral structure were known. All of the stages of viral replication had been outlined, the viral genes had been mapped, and their functions deduced. Sanger was elaborating the complete nucleotide sequence of the viral DNA. Kornberg was using ϕX to disentangle the complex enzymology of DNA replication. Many details were still obscure, such as the manner and order of assembly of the progeny virus particles and the enzymology of lysis of the host cells. But these seemed likely to be of parochial interest, specific to this not especially important virus. Graduate students, perceptive of future scientific opportunities, were choosing to work in other laboratories.

It was time to initiate another research program. A major shift, as into neurobiology, was alluring but would require a few years to learn quite different techniques, acquire a background in the field, and establish a quite different laboratory. I had now taken on a variety of outside interests and commitments that would severely conflict with the concentrated effort required to establish a wholly new research program. And I believed these interests merited a significant share of my time. After some thought, I therefore decided on a research direction that would make good use of my current skills and established laboratory. At this stage, I also sought a problem closer to practical application for societal benefit. I chose the field of nitrogen fixation.

All living forms require nitrogen, usually as ammonium. The nitrogen comes either from consumption of other organisms or their decay products or, for plants and many microorganisms, by fixation of nitrogen from the atmosphere. Plants cannot fix atmospheric nitrogen themselves but rely on microorganisms in the soil with which they establish a symbiotic relationship. The nitrogen thus supplied is frequently limiting for growth. Crop yields are markedly improved by the application of costly nitrogen fertilizer, produced by industrial processes but coming ultimately from the atmosphere.

The biochemistry and genetics of nitrogen fixation by those microorganisms with that capacity was poorly known and seemed a ripe subject for the application of molecular biology. And the possible benefits to agriculture of improvement in nitrogen fixation techniques were evident. Because this research would primarily involve microorganisms, many of our well-established techniques and laboratory facilities would be immediately applicable, though new modes of assay and equipment for work under anaerobic conditions would have to be added.

Considerable thought was given to the design of experiments to approach this problem. The first steps—acquisition of appropriate bacterial strains and familiarization with assay techniques—were underway when my career took another distinct turn.

1. With mother, Rose, and brothers, Allen, Jr. (second from left), and Richard (right), 1938.

2. Laboratory at Iowa State, 1951.

3. Biology faculty and research associates at Caltech, 1958. Left to right: Max Delbrück, Ernest Anderson, Norman Horowitz, George Laties, Roger Sperry, Harry Rubin, Alfred H. Sturtevant, George Beadle, Sterling Emerson, Henry Hellmers, Robert Sinsheimer, Edward Lewis, Geoffrey Keighley. Absent: James Bonner, Henry Borsook, Herschel Mitchell. (Courtesy California Institute of Technology)

4. Speaking at dedication of Arnold and Mabel Beckman Laboratory of Behavioral Biology, 1974. Seated, left to right: Arnold Beckman, Mabel Beckman, Harold Brown. (Courtesy California Institute of Technology)

5. Inauguration as chancellor of UC Santa Cruz by UC president David Saxon, 1978. Photo by B. Lee. (Courtesy UC Santa Cruz Photo Services)

6. With President Emeritus Clark Kerr at dedication of Clark Kerr Hall on Santa Cruz campus, 1978.

7. With Dalai Lama at University House, UC Santa Cruz. Photo by Carol A. Foote. (Courtesy UC Santa Cruz Photo Services)

8. President Saxon and the chancellors, 1982. Left to right: Daniel Aldrich, Robert Huttenback, Julius Krevans, Robert Sinsheimer, James Meyer, David Saxon, Tomas Rivera, Richard Atkinson, Charles Young, Ira Michael Heyman. Photo by Saxon Donnelly.

9. At retirement dinner, receiving "chair" symbolic of donated Sinsheimer professorial chair in molecular biology, 1987. Photo by Don Fukuda. (Courtesy UC Santa Cruz Photo Services)

10. Speaking at dedication of Sinsheimer Laboratory, 1990. Standing, front row, left to right: David Gardner, Libby Gardner, Karen Sinsheimer, Cliff Poodry, Kathy Stevens. (Courtesy UC Santa Cruz Photo Services)

11. At seventieth birthday symposium, gathering of former students and postdoctoral fellows, UC Santa Barbara, 1990. *Front row, left to right:* John Abelson, Graham Darby, Arnie Levine, Larry Dumas, John Newbold, Walter Fiers, Bob Sinsheimer, Bob Nutter, Paul Johnson, Lee Compton, Lloyd Smith. *Center:* Ali Szalay, Jim Strauss, Marshall Edgell, Ellen Strauss, Tohru Komano, Nicole Truffaut, Jane Cramer Harris, Jim Koerner, Moises Eisenberg, Dave Denhardt, Tony Zuccarelli, Bob Rohwer, Galina Moller. *Back:* John Hall, Erica Wickstrom, Harry Beevers, John Sedat, Phil Hanawalt, Harry Noller, Björn Lindquist, Ted Young, Clyde Hutchison, Nigel Godson, Stan Krane, Alice Burton, Peter Baas, Regis Kelly, Anne Haywood, Lorrie Greenlee, John Kiger, Helmut Steiger, Karel Grohmann, Barklie Clements, Akio Fukuda, Jean Paul Revel.

19

Transition 5

"David Saxon is calling you from the president's office at the University of California."

David Saxon, a physicist, had been a classmate at MIT. I particularly remembered him in the atomic physics course. I had encountered him at UCLA occasionally in the intervening years and had noted with interest his selection as president of the University of California in 1975. Why was David calling me on a rainy February morning in 1977? Simply put, he was looking for a chancellor to head the University of California campus at Santa Cruz, the newest campus in the UC system, and my name had been put forward. Would I be willing to consider this position, to meet on the campus with the search committee and then with himself and other top university officials? I would think about it.

I knew a little about Santa Cruz. My daughter Kathy had been a student there for three years, 1968 to 71, and I had visited her on several occasions. It was an "experimental" campus, distinctive in the UC system for its collegiate structure. It sought to relieve the anonymity and alienation of students on the large UC campuses by clustering them into living and teaching units, the colleges, each with some six to eight hundred undergraduates. Physically, it was extraordinarily beautiful, located in a redwood forest at the edge of meadows overlooking Monterey Bay, seventy-five miles south of San Francisco.

The invitation created an intense personal dilemma. I was completing my twentieth year at Caltech—and I loved the institute, its scientific intensity, its freedom from bureaucracy, the wealth of able colleagues in all of science. I had long since decided I would not trouble to consider offers from other academic institutions. But, in terms of career, I was at a moment of transition. My ϕX research was phasing out and I was preparing for a quite different research program in nitrogen fixation. I was in my ninth year as chairman of biology and I had informed the president that I felt my tenth year should be the last. I wanted to devote more time to research and scholarship and a fresh chairman with fresh ideas would be salutary for the division.

In addition, I had become increasingly concerned with the dramatic potential and inevitable social consequences of the advances in biology. My experience in the "recombinant DNA" controversy and my inability to move Caltech even modestly toward a broader intellectual base led me to believe that the impetus to consider seriously such issues would not come from within science. Perhaps a new and (as I thought) growing, yet unshaped, university such as Santa Cruz would provide the medium in which my concerns could find expression.

At fifty-seven, I could make but one more major career move if I so desired. So, yes, I would visit Santa Cruz, with no commitment asked nor given.

The campus was as beautiful as ever, with sunlit paths, shady ravines, and towering redwoods looming against the sky. But the physical plant, only 12 years old, looked a bit shabby—walkways were crumbling, smudged or scraped walls needed paint. The campus was clearly troubled. The founding chancellor, Dean McHenry, had retired in 1974. His successor, a relatively young professor from Berkeley, had proven inadequate to the complexities of Santa Cruz and had been forced to resign after a year and a half. The present chancellor, Angus Taylor, was a UC veteran sent down from the central administration to fill in until a new chancellor could be selected. At that time, the causes of the disarray and discontent seemed to me either petty or obscure. They seemed manageable. As I was later to learn, however, the pettiness concealed deep ideological differences and the obscurity derived from basic images and policies deeply embedded in the University of California, but then unknown to me.

I met with the chancellor, faculty, staff, and, somewhat to my surprise, students. Their questions were not very probing. All seemed to recognize that the present situation was untenable, but there was no clear vision of the future. They were looking for a savior to lead them—acceptably—out of their sea of troubles. The most prescient question came from an intermediate level administrator who asked: "You have clearly been a very successful scholar. Why do you want to become an administrator?" I did not realize then that, in the UC system, the two were incompatible. I went on to meet with David Saxon and some of his central UC administration. David indicated his full support for Santa Cruz. In retrospect, I don't think he had a grasp of the depth of its problems.

I returned to Caltech uncertain whether I would accept were the position offered. A week later, it was. I was their first choice.

My career had encompassed the two finest institutes of technology in the nation, as well as a period at one of its finest agricultural schools. Now I was asked to take charge of a relatively new and (I thought) growing university to guide its destiny into maturation. Little did I know that it was in fact moribund and regarded by many as a misbegotten child of the UC system in dire need of resuscitation. What tipped the balance in my mind? I believe it was the challenge and, as I then thought, the opportunity. I could stay on at Caltech, comfortably and happily, another dozen years in a predictable milieu. Or I could essay a very different role in a very different environment, with a different context and horizons.

I accepted. And indeed it turned out to be different—from my expectations, from past experience—and a tremendous learning opportunity.

UC Santa Cruz:
The Anomalous Campus

20

The Prologue

The University of California at Santa Cruz was conceived as an alternative to the UC mold, the Berkeley model. Unfortunately, no one was willing to pay for it.

And in the critical moments of its early years, it was triply betrayed, initially by the critically flawed logic of its founders and subsequently by the university administration that had created it and the community that had solicited and welcomed it. Both latter betrayals reflected the cold-blooded consequence of political change in public institutions in which officials assume no responsibility for the pledges and plans of their predecessors.

Within the UC system of campuses, Berkeley and UCLA had been the paradigms. Large—with thirty thousand students each—impersonal, emphasizing professional schools and research institutes, and boasting intense semiprofessional athletic programs, they were by their own standards highly successful. But for many undergraduates these campuses engendered feelings of alienation, a loss of individuality, a sense that they were ignored in favor of more advanced training and research.

In the late 1950s and early 1960s, under the aegis of the California master plan for higher education, the University of California initiated the development of three new campuses, two in the southern part of the state (now UC Irvine and UC San Diego) and one near the San Francisco Bay area (now UC Santa Cruz), the last intended in part to relieve enrollment pressure on Berkeley. In what Peter Hall in his 1982

163

book *Great Planning Disasters* calls a "near disaster," these campuses were launched under the assumption that UC enrollment would increase by 135 percent between 1960 and 1975 and continue to rise thereafter. However, demography failed. The birth rate fell markedly throughout the 1960s and 1970s. Net migration into California in the early 1970s was one tenth the rate of the 1960s. Projected forward, these statistics predicted a decline in university enrollments in the 1980s.

At the same time, in the late 1960s and throughout the 1970s, the university experienced grievous fiscal problems. Costs continued to rise, but the student radicalism of the period, particularly at Berkeley, destroyed the university's prestige and alienated legislative and public support. Governor Reagan, after his election in 1966, immediately led the board of regents to fire President Clark Kerr and, subsequently, repeatedly reduced the university budget. His successor, Jerry Brown, who seemed to have a personal vendetta against the university, was little better. To quote Peter Hall: "Thus from 1969 onwards, campus planning was dominated by extreme caution. The main reason for this was the hard facts of demography, but undoubtedly the bleak story of finance came into it, and reinforcing that the loss of the University administrators' self-confidence."

In this climate, the newly launched campuses suffered, and UCSC the most grievously. UC Irvine consciously set out to become another Berkeley. UC San Diego set out to become a powerful research and graduate-oriented institution, focused especially on the natural sciences, as much like Caltech as possible in a public university. But Santa Cruz was set on a different path.

Clark Kerr had been a student at Swarthmore in the early years of Frank Aydelotte's presidency there and acquired a high regard for the worth of the education that can be provided by a small liberal arts college. He was also aware of the limitations of such institutions, particularly with respect to the scholarly facilities available in the large universities—the libraries, science laboratories, arts studios, concert halls, computers, and so on. Kerr and Dean McHenry, a former graduate school classmate at UCLA who became the founding chancellor at UC Santa Cruz, were also concerned by the increasing emphasis on departmental structure and prestige at the large campuses. This led to the fragmentation of knowledge, the inability to fashion a coherent curriculum, and an increase in professionalism and focus on research (as contrasted to teaching) within each department. In addition, Kerr recognized that large, impersonal campuses such as Berkeley engendered in many undergraduates a dislike for what they came to perceive as a repressive administrative bureaucracy rather than a partner in learning.

To counter these tendencies, Kerr and McHenry conceived of a different UC campus. In its design, they borrowed from a number of more-or-less distinctive ideas floating about in U.S. higher education as well as from some of the aura of Cambridge and Oxford. They conceived of the campus as a cluster of small (about 750 students) undergraduate, residential liberal arts colleges. Since the ultimate enrollment was to be 27,500 students, one plan called for twenty-three such colleges, together with graduate and professional schools. As the concept envisioned mostly small classes, most classrooms were to be located within the colleges. To ensure close faculty interaction with the students, most faculty offices (save for the science faculty) would be in the colleges as well. Indeed, it was hoped that some junior faculty would live in each college, as well as the college head, the provost. As a public university, students could not be required to live in a college, but it was expected that the great majority would choose to do so.

It is not clear that Clark Kerr's vision extended beyond a literal conglomerate of Swarthmores (as he remembered it) sharing such common facilities as a library, athletic fields, and financial record keeping. It was not even clear initially whether the shared facilities were to extend to science laboratories or art studios, although inevitably for fiscal reasons these were centralized. It seems odd that Kerr did not recognize the salient and critical distinctions between a Swarthmore and a UC campus. Swarthmore provided a student/faculty ratio of 9/1; for a UC campus today the ratio is closer to 20/1. Also, the systemwide standards and criteria for selection and advancement of faculty at a research university such as UC are necessarily quite different from those appropriate to a Swarthmore.

When Dean McHenry approached Kenneth Thimann—a distinguished botanist at Harvard who had been quite active in the Harvard residential house system—to help launch the science program at UCSC and to become the founding provost of the third college, McHenry suggested that the science faculty should exclusively teach during the academic year and only undertake research during the summer! Thimann (and others, notably Francis Clauser) pointed out that this was a completely impractical way to do research and that the campus could never attract good scientists with such a program. Then McHenry wanted to begin the campus without any graduate students, at least for some years. Again, Thimann and others had to explain that science faculty expected to work with graduate students in their laboratories and that recruitment of good faculty would be impossible without provision for these. It is remarkable that McHenry, a senior university administrator, could have been so näive about the needs of scientists.

Even more curious, Kerr seems to have overlooked the obvious fact that the more he emphasized undergraduate education at Santa Cruz, the more he would blur the critical distinction of the California master plan for higher education, which reserved doctorate and professional education to UC and made UC the research arm of the state, while expecting the California State University system to provide the bulk of four-year undergraduate education. Even worse, in accord with the master plan and deeply embedded in the UC psyche is the notion that the CSU colleges have an inferior role in the educational hierarchy. For a UC campus to emulate that role inevitably gave that campus an inferior status within the UC system, long dominated by Berkeley and UCLA alumni. And legislators would look askance at a UC campus performing a function similar to that of a CSU campus but at much higher cost per student.

All of these inherent contradictions came to pass and together with the grim external pressures enmeshed the campus in a strangling coil.

The choice of a site for the new campus reflected the romantic—or as Dean Tschirgl of Berkeley called it, the "nostalgic"—vision embodied in its plan. After extensive review, two sites were in contention: one in the Almaden Valley just south of San Jose, the other on the Cowell Ranch in the redwoods just outside of Santa Cruz, overlooking Monterey Bay. Both communities eagerly sought the new campus. After a visit to both sites on a warm summer day, the regents, taken by its beauty and perhaps its romantic ivory-tower isolation, chose Santa Cruz.

It is of interest that Governor "Pat" Brown favored Almaden. He believed it important that a university be in an urban center for student access, for campus access to external sources of support, and for access to cooperative endeavors in the sciences and the arts. Additionally, in this age of two-career families, an urban locale provided much greater opportunity to satisfy both career needs and thereby facilitated faculty recruitment. The governor was correct.

In fairness, I should note that the Cowell Ranch property could be purchased in one transaction at a favorable price from the Cowell Foundation, a frequent donor to UC, while the Almaden property would have involved some seventy parcels and required complex negotiations. Also, early concepts of the Santa Cruz campus envisioned a significant community developing around it and along the adjacent coast. This vision did not anticipate the rise of the local "no-growth" movement.

UCSC opened with one college (Cowell) in 1965 and added a college a year for the next six years. In the mid-1960s, the collegiate concept proved very popular with potential UC students. In addition, the disturbances at Berkeley led many parents to steer their children elsewhere. Santa Cruz was still "in the neighborhood." And so for a few years, entering Santa Cruz freshman had the highest SAT scores of those at any UC campus.

Starting such a new and different campus is a heady, once-in-a-lifetime venture and the first faculty and administrators poured their energies into its establishment—at the cost, sometimes unrecognized, of their scholarship. Because faculty had to be recruited before the students could be admitted, the campus started off with a student/faculty ratio of 10/1. With this richness and intense faculty involvement, the vision of small classes and close student/faculty interaction could be realized—for a time.

Without much UC precedent, the organization of the colleges had to be invented by "doing," by trial and error. It is significant that three of the leading figures during the establishment of the early colleges had had experience at Oxford or Cambridge and used these as models, without much thought of the difficulty of transplanting five-hundred-year-old British traditions to the effervescent environment of California. The English concept of faculty living and eating in close attachment with their colleges soon faded before the more American attachment to family, especially in a small-town environment. Moreover, with the establishment of collegiate organizations and the accompanying staff to provide students with personal attention and services such as housing coordinators, activities directors, psychological counselors, and so on, it became apparent that these liberal arts niceties were costly. And whereas at Oxford and Cambridge the colleges had their own substantial endowments, at Santa Cruz they did not. The funds had to come from the campus's educational allotment from the central UC administration. Essentially formulaic, this took no account of the special requirements inherent in the Santa Cruz design.

As the campus grew, the student/faculty ratio steadily rose toward the UC norm, which in turn, given the hard fiscal times, was also steadily rising.

As soon as there was more than one college, some structure was needed to coordinate the curricular offerings within the various disciplines. Antipathetic to departmental structures and disciplinary boundaries, Dean McHenry named the disciplinary units "boards of study," not "departments." Since the campus was planned to grow to 27,500 students, with fifteen hundred or more faculty members, no need was felt to concentrate faculty in certain disciplines or subdisciplines to create centers of strength. To the contrary, some twenty-three boards of study were established, all thinly staffed and each diverse. Because no thought was given to the possibility of arrested growth, the faculty was stretched very thinly.

Conceptually, the boards of study were to be merely coordinating bodies. But, inevitably, the demands of disciplinary education and the need for provision of an appropriate range and depth of courses within each field began to be felt. Although a faculty member nominally held a 50 percent appointment in a board and a 50 percent appointment in a college, educational control gradually but inexorably shifted to the boards. Disciplinary pressures increased within the boards and courses given by faculty within the colleges became for the most part increasingly peripheral and "Mickey Mouse," and their requirement increasingly resented by many faculty members. And, inevitably, since the boards and colleges competed for the same always inadequate pool of funds, the tension between these two organizational sets, with their nearly orthogonal missions, increased.

In theory, the campus was a matrix or two-dimensional structure. Such organizations can work well in large hierarchical enterprises with an externally imposed mission. Thus, in the Radiation Laboratory at MIT, the operational divisions of land, sea, and airborne radar could call on the expertise of the component divisions that designed transmitters, receivers, antennas, display units, and so on. Similarly, at the Jet Propulsion Laboratory of Caltech, the mission-oriented groups (e.g., Mariner, Voyager) could call on the technique-focused groups for design and operational help. It is not at all evident, however, that such a matrix organization can generate its own missions, particularly within the democratic ethos of academia.

The college-board structure was supposed to produce "creative tension"; instead it produced deadlock. This became particularly manifest in the recruitment and promotion of faculty.

Each college sought to have a broad "theme" to make it distinctive from the others, but one that could encompass a wide spectrum of faculty and be compatible with a "liberal arts" orientation. Cowell, the first, chose humanities and Western civilization; Stevenson, the second (named for Adlai Stevenson), chose the social sciences, with a focus of political science. Crown, under Kenneth Thimann, then chose the natural sciences. After these, choices were harder. Merrill vacillated between international affairs and Third World problems. Porter, the fifth, became the fine arts college.

Kresge, the sixth, began as the physical sciences college, then switched to ecology. Its first provost, Robert Edgar, a distinguished geneticist from Caltech, became enamored of the burgeoning "sensitivity" movement of the time and sought to build a college based on the principles of "sensitivity training" and "T groups." The idea seemed to be that if all of the participants in the college could be sensitized to each other's needs, backgrounds, and aspirations, then, being of good will, they could evolve programs and curricula to help each other learn and grow. As might have been expected, the implementation of such a concept involved endless discussions and wrangling, and failure to achieve consensus in the end produced paralysis. Kresge College became notorious for its feeble academics and rather libertine lifestyle and its sorry reputation pulled down that of the entire campus. Edgar resigned leaving the college in a state of stagnation.

The seventh college, Oakes, was dominated by Herman Blake, a charismatic black sociologist who sought to make it a port of entry for minority students into higher education. By providing special tutoring services, funded with external grants, for minority freshmen and sophomores, who often came with inferior high school backgrounds, he sought to "bring them up to speed" so that they would be ready for the regular academic program in the junior year. This program was imaginative and, in good degree, successful, but it had the unfortunate effect of creating a minority "ghetto" at Oakes to which most of the minority students then gravitated, even if they were initially assigned to other colleges.

College Eight, as it was called, although established in 1973, had no buildings until 1989 and was thus a commuter college. Its theme was environmental studies.

With their "themes," the colleges had their own recruitment agendas. But each faculty appointment had to be in a board as well, which had *its* recruitment agenda. Frequently, these requirements were in conflict, for which there were three possible solutions. In times of rapid growth, two appointments could be made available, one to satisfy each agenda. As growth slowed, this option became unavailable. Or sometimes the board's first choice and the college's first choice could be bypassed in favor of a second-choice candidate, probably inferior to either first choice but more broadly acceptable. Peace was preserved, but at a price. Or, finally, the issue could be passed up to the central campus administration for resolution, which left one side happy and the other bitter.

Faculty promotions and particularly tenure reviews brought these conflicts to a head. The boards and the colleges simply had different missions and correspondingly different criteria. The board valued scholarship, research, and professional teaching and was generally in accord with the systemwide UC standards. The college valued service in the college, counseling and working with students, teaching in interdisciplinary college courses, and participation in "college building" (especially in the early years). The college functioned more like a club. Junior faculty, who often had been encouraged to devote their energies to collegiate affairs at the expense of their research and scholarship, now found themselves caught.

Tenure decisions for a member of the faculty were made in both one's board and one's college. Understandably, these frequently diverged. Indeed, individual senior faculty, members of both the board and the particular college, were known to vote oppositely in the two circumstances using different criteria. The divergent votes would then be reviewed by the Academic Senate Committee on Academic Personnel, a six-person body and, depending on their inclinations preferentially to value board or college service, they would frequently produce a split vote. Which left the decision to the central campus administration—that is, the chancellor.

Most often, the chancellor tended to uphold the disciplinary or board standard as most consistent with broad UC standards, from which Santa Cruz had never been excepted. Then, as such negative tenure decisions most often resulted in the departure of an often well-liked member of a college, faculty and students in the college were outraged. Thus, "creative tension" became a constant irritant, a boil on the campus ambience. Page Smith, the founding provost of Cowell College, resigned in a fury over a negative tenure decision for one of his protégés.

In addition to its primary "theme," the faculty of each college was supposed to be able to provide a liberal arts education within each college. While students were free to take courses in any college or board, it was expected that they would take most of their courses within their home colleges. However, since the faculty offices, save for the scientists, were in the colleges, this policy resulted in the dispersion of the faculty of any one discipline over several colleges. The natural result was a loss of intellectual vitality in the disciplines. An economist, who might be the only one in his or her college, had limited interaction with his or her fellow economists and, if a junior faculty member, had limited opportunities for mentoring and introduction to the profession. Several younger faculty members were professionally destroyed by this unintended but unfortunate side effect of the initial college concept.

Starting in 1971, with the return of calm to Berkeley and some disillusion with the development of the campus, freshman applications to Santa Cruz began to decline, as did the academic ability of applicants. Other factors hastened the decline. As one of his idiosyncrasies, Page Smith was resolutely opposed to academic judgments. In the first year, he had pushed through a "narrative evaluation" proposal to replace letter grades, in addition to a Pass/No Record plan, which meant that no failures were ever recorded. In principle, the narrative evaluation provided, instead of a categorical letter grade (A, B, C, D), a written paragraph or two describing and evaluating the student's performance in the class. In theory, it provided the possibility of a more multi-dimensional evaluation than a single averaged grade. Also, it reduced competition among students simply for grades. In small classes, where the instructor can get to know each student and has lighter teaching duties, the narrative evaluation can work well. However, unfortunately, as the student/faculty ratio rose, the possibility of meaningful evaluation diminished. Many faculty resorted to various assortments of stock phrases, even computerized ones.

In addition, applicants to graduate or professional schools were penalized. Admissions committees at the better schools, often deluged with applications, were simply not willing to read through pages of text to evaluate a few Santa Cruz students. While personal intervention in the form of communications from Santa Cruz faculty to colleagues at other schools could often bypass this block, at the professional schools of medicine or law the net effect was to place much greater emphasis on the MCAT or LSAT test scores for Santa Cruz students.

Also, to the general public, the narrative evaluation—or "no-grade" scoring, as it was popularly perceived—was seen as an indication of academic laxity and low standards, and this deterred many prospective students.

The student unrest of the 1960s did not bypass Santa Cruz. It lacked the critical mass of students necessary to produce riots such as those at Berkeley, but other radical elements of the time, such as politics and drugs, swept the campus. As a new campus, Santa Cruz lacked the traditions and customs that helped most older campuses to avoid extremes and maintain a reasonably orderly structure. Santa Cruz bent sharply with the winds. When Clark Kerr went to address the first four-year graduating class at commencement in 1969, he was forced to sit through an hour of diatribe and then was never allowed to speak. At the regents' meeting at Santa Cruz in 1968, the students surrounded and blockaded the bus containing Governor Reagan and the regents after the meeting, creating a tense situation. One outcome was the cessation of regents' meetings on campuses for more than a decade.

This "radical" image combined with the soft "no-grade" image to create a vision of Santa Cruz as a far-out, laid-back campus where one went to flake out and smoke pot under the redwoods, not a serious academic locale. While doubtless attractive to some, this image further deterred many more prospective students.

And then, in the early 1970s, the Santa Cruz community was the scene of several shocking murders, two involving a campus employee. Santa Cruz became known as the "murder capital," and enrollment applications plummeted. Originally, there had been a surplus of applications from freshmen and transfer students. By accepting more transfer applicants, a decline in enrollment could be avoided for a time, but these too soon declined. Enrollment leveled off in the mid-1970s and was predictably about to decline.

From the beginning, it was evident that the Santa Cruz vision of college education would cost more than the Berkeley pattern. Economy was, in fact, a primary reason for the large campus model. This question was raised repeatedly by various regents in early discussions of the Santa Cruz concept. How could personalized attention and guidance in the colleges, education in small classes, and all of the other attributes of small liberal arts college ambience be conducted at the same cost per student as the large impersonal lecture classes of a Berkeley or UCLA? Kerr and McHenry repeatedly waived aside such concerns with bland pronouncements that all could be worked out. But in the final approval given by the regents to the Santa Cruz Long Range Development Plan was included the proviso that the plan was "subject to the condition that State-funded capital and operating costs shall not exceed the comparable costs per student on other campuses."

For a time, the enthusiasm and dedication of the newly recruited faculty on the new campus could carry the extra burden. And initially, the surplus of faculty accompanying a build-up phase helped. But as enrollment growth slowed, this surplus vanished. And faculty either "burned out" or, by inclination or necessity (to maintain their status or to achieve tenure), returned to their more professional and scholarly pursuits.

The systemwide UC administration was not helpful. After Kerr's dismissal, it seems clear that much of the central administration looked askance at this strange organism in their midst. Most of the administrators were products of Berkeley or UCLA—the UC system is highly inbred. The Santa Cruz concept implicitly suggested flaws in the education they had received. They expected, perhaps even hoped, this aberration would fail. So where was the required extra funding to come from? Conceivably, as at Cambridge, from college endowments. Most of the colleges were in fact endowed in exchange for the name, but the endowment funds were mostly spent for college facilities that the state would never fund, such as a provost's residence, a college library, commons rooms, and so on. Very little was left to provide endowment for ongoing expenses.

I estimated, much later, that a 25 percent augmentation of the Santa Cruz budget above that provided by UC formulas would have provided the funding required for a minimally satisfactory collegiate operation. In the first decade, this sum would have been less than 1 percent of the UC state budget. It is possible that Kerr thought he could manage such a small diversion out of the UC budget.

But the new president, Charles Hitch, provided no special funding. His inclination was clearly shown as he gave UC San Diego—building at the graduate level and emphasizing science and the professions—a disproportionate share of the new higher-level academic positions available, while providing Santa Cruz almost exclusively junior-level positions.

The unkindest cut was soon to fall.

The emphasis on building a new college every year consumed the energies of the campus. While it was recognized in the initial plan that Santa Cruz as a UC campus would in time have significant graduate programs and professional schools (engineering, business, forestry, and landscape architecture were mentioned), these were, with few exceptions, postponed. In part, this postponement reflected indecision as to how graduate and professional education could be well integrated with the undergraduate liberal arts programs of the colleges.

Graduate programs were, however, begun primarily in the natural sciences where the faculty needed graduate students to assist in their research. And engineering was initiated. Francis Clauser, a distinguished aeronautical engineer and Caltech Ph.D., was brought to be the first dean of engineering. Projections were made that by 1975, UCSC would graduate sixty B.S., fifty M.S., and sixty Ph.D. degrees in engineering per year.

At this time, the California Coordinating Council for Higher Education commissioned Frederick Terman, then dean of engineering at Stanford, to perform a study of engineering education in California that was completed in the spring of 1968. Terman performed a thorough engineering analysis of the situation, which led him quickly to conclude that low (or reasonable) cost and high quality of engineering education were linked to adequate sizes of undergraduate and graduate instructional programs: "When the degree output is small, the instruction cost index [direct instruction cost per student credit hour] increases very substantially."

From this observation, he concluded the following:

Programs producing more than 200 B.S. degrees per year are more attractive and of higher quality. . . . In California more undergraduate engineering programs than are really needed are competing for a pool of undergraduate students that is now too small to go around. This situation will continue for at least a decade. . . . No California institution is currently overloaded by more M.S. students than it wishes to handle. In fact, only four of the sixteen institutions that awarded M.S. degrees in 1966–67 have M.S. programs that achieve the minimum desirable size. . . . Many California institutions would have stronger engineering programs if they gave up some of the fields of engineering in which their enrollments are now very small.

Terman was particularly critical of the UC plans to establish engineering programs at all eight general campuses:

Because underpopulated engineering programs are expensive to operate and tend to be simultaneously of poorer quality than programs of more nearly optimal size, new engineering programs should have been established only as fast as already existing programs were becoming filled up. . . . The soundness of the new graduate engineering programs being established at various University campuses will to a considerable degree be tied to the success of the associated undergraduate engineering programs. As an undergraduate program builds up to 200–300 bachelor's degrees per year, the associated graduate program has the potential for being good enough, large enough, and sufficiently diverse to be simultaneously economic in operation and attractive to students.

As a result, at least two or three of the . . . new engineering programs on University campuses can for many years be expected to lag conspicuously behind in the competition for undergraduate students. . . . The University would be serving the State best if it would concentrate all resources available for full-time-on-campus graduate education in engineering at four or at most five campuses.

In consequence of this report, the coordinating council recommended to the university that the two most recent engineering programs at Santa Cruz and Riverside be terminated. The engineering council of the university vigorously rebutted this conclusion, arguing for the academic value of engineering programs as well as their service to the economy of the state. UC President Hitch, in the spring of 1968, resolved the issue by abruptly cancelling the two engineering programs. He did so, apparently, primarily for fiscal reasons. Confronted with restricted state budgets under Governor Reagan, he concluded that these new engineering programs would be quite costly during their developmental phases, and they could be eliminated.

It was an accountant's decision, but it took no account of its effect on the morale of the young campus, the subsequent composition of the faculty and the student body, the attractiveness of UCSC to prospective students, and its status in the UC system. It took no account of its effect on the ability of the campus to establish meaningful ties to the logical nearby source of external support, Silicon Valley. It left the campus unbalanced and highly vulnerable to the periodic swings of student interests. The decision was woefully short-sighted and a disaster for the three-year-old campus.

As the budgetary crunch of the Reagan years took hold, no effort was made to shelter the vulnerable newer campuses. Indeed, small and lacking economies of scale, they suffered the more. But no reallocation, even partial, of the systemwide resources could be considered. The two large campuses, Berkeley and UCLA, which had for historical reasons the great bulk of the resources, were the "flagships" of the system. Their eminence and prestige were to be maintained regardless of the cost to the rest of the system. "Every tub on its own bottom" became the maxim. It could be argued that maintaining the prestige of Berkeley and UCLA encouraged the state to be more generous to the UC system than would otherwise have been the case. The validity of this, as with other "trickle-down" theories, is difficult to verify.

The Santa Cruz community that had eagerly solicited the presence of UC in the late 1950s and early 1960s was a somewhat drowsy seaside town; its population in 1960 was 25,596. At the beginning of the twentieth century, it had been a lively source of lumber, limestone, and tannin, much of which had come from the ten-thousand-acre Cowell

Ranch. But, as this industry faded, it had become on the one hand a retirement community and on the other a summertime seaside resort for the San Francisco Bay Area, with a beach amusement park and allied businesses. The owners of the summer rental units saw an influx of students as potential wintertime tenants. The business owners, of course, saw enhanced business and the possibility of the university as a catalyst to bring about the development of the city and adjacent coastal areas.

But three major changes in population and a strong shift in public sentiment changed all of this. For one, as the 1960s tide of hippies and flower children ebbed in the Bay Area, it left its debris on the beach at Santa Cruz. Late into the 1970s and early 1980s, parts of Santa Cruz could be described as "museums of the hippies." Some of the townsfolk blamed this influx on the university, but there was little obvious cause and effect. Second, the explosive growth of Silicon Valley and San Jose drove many people, dismayed by the sudden urbanization of a formerly semirural area, over the ridge to Santa Cruz, where they were determined that the same growth should not recur. With them also came a flock of Silicon Valley employees, who used Santa Cruz as a quiet bedroom community from which they could commute "over the hill."

And third, of course, was the influx of the university population. Previously unanticipated, a California Supreme Court decision in 1971 gave students the right to vote in their campus communities. The judges who made that decision probably never considered its impact in small university towns. The introduction of the university population, predominantly students, into Santa Cruz fractured and fragmented the community. In many ways, it is no longer a community. Students form the largest and therefore the decisive voting block, but they have no lasting stake in the community. Local politicians have pandered to student predilections toward predictable short-term ends.

The students at Santa Cruz come mostly from urban and suburban areas to the relatively open space of Santa Cruz where, oblivious to the developments that made possible their presence there, they are easily persuaded that not another tree should be cut nor a field built on. Indeed, after the first two colleges were built, the students have repeatedly protested the building of each new college, sometimes violently. The inherent contradiction between their beliefs in universal education (especially for minorities), their admiration for the Santa Cruz college system, and their resistance to expansion of the campus to provide facilities for more students like themselves is lost on them. Environmentalism is often a cover for self-interest.

The principal issues in the local elections of the early 1970s were proposals to build a sizable community of about ten thousand people on the coastal lands north of Santa Cruz and to build a hotel adjacent to a popular coastal area, Lighthouse Point, in Santa Cruz itself. Mobilizing the student vote against that of the business community and the old establishment, the no-growth politicians carried the day. Building on this, they subsequently defeated proposals by the Department of Highways to build a freeway over the hill from San Jose, which was to have tied into a six-lane road, promised by the county, to the campus. This would have provided excellent access to the campus, which today can be reached only by two-lane roads passing through residential areas. (The access commitments pledged by city and county have since been conveniently forgotten and burked.) Buoyed by their success, the local politicians sought to curb the growth of the campus, initially envisioned to have 27,500 students on its two thousand acres, to its then five thousand students.

By the mid-1970s, then, the campus lay stagnant, its enrollment about to decline, its initial energies exhausted, its academic structure unworkable, its educational reputation in tatters. A sorry tale of inept planning and administrative neglect.

In June 1973, Dean McHenry went to the regents to request that the campus plan be scaled back, for at least a decade, to a maximum of seventy-five hundred students. He recognized that such a decision meant that no significant graduate nor any professional programs could be developed over that time. The campus would limp along in its unfinished, truncated state, but such a decision would bring peace with the community—and perhaps foreclose some of the ceaseless debate on campus. Four months later, McHenry announced his intention to retire.

Academic reformers, thirsty for the opportunity of innovation, hailed the advent of Santa Cruz, almost willfully overlooking its inherent contradictions. Even as the campus paid the price and patently floundered, they could not acknowledge their wishful thinking and oversights. Thus in 1978, Gerald Grant and David Riesman could write in *The Perpetual Dream*:

When it was aborning, few campuses seemed to offer as much promise or to draw as much praise as the University of California at Santa Cruz. Yet a decade later, it was demoralized. . . . Enrollment at what was once the most favored campus in the California system was falling off sharply. A once-buoyant faculty was feeling middle-aged, and some feared that the great white whale of the counterculture was rotting high on the beach at Santa Cruz. . . .

Yet we hope to show in this account that such a portrait, while true enough in outline, overstates the gloom. Despite its recent reverses, Santa Cruz was, and in most ways continues to be, one of the most genuine successes of the last decade. One learns this not from its dispirited faculty but from the students who have been drawn to the stunningly beautiful campus.

Oh, Pollyanna!

Thus, UC Santa Cruz began its existence burdened with an incoherent academic plan and an insolvent fiscal budget and set in an isolated, insular community. But of most of this I was blissfully unaware.

21

The Education of
a Chancellor

"President Saxon, Governor Brown is on the phone."

The nine chancellors of the UC campuses form the council of chancellors. We met at least once a month (and, frequently, more often) with the president and vice-presidents of the university to discuss subjects of common interest such as budgetary matters, affirmative action, legal concerns, and policies affecting faculty, staff, or students. As my "spring training," I was invited to attend the council of chancellors meeting in June 1977 prior to my actually assuming the chancellorial position at Santa Cruz.

I was surprised and disconcerted by this meeting. The discussion was very much "nuts and bolts"—concerned with quite specific details of management rather than with broad issues of educational or administrative policies. (I had thought chancellors were to be concerned with educational policy.) And here was the president of the university engaged in an hour's discussion with the governor of the state over (as I learned) specifics of the university budget, which was then on the governor's desk. I had thought the UC was insulated from the political scene through the mediation of the board of regents; I quickly learned that this was seldom true. This meeting was a foretaste of the direct vulnerability of the university to the fiscal conditions, political personalities, and shifting policies of the state that I would later encounter in abundance.

The bright sunlight slanted through the tops of the surrounding redwoods into the old quarry, now converted into a natural amphitheater.

"I, David Saxon, president of the University of California, by virtue of the authority vested in me by the regents of the university, do hereby inaugurate you, Robert L. Sinsheimer, as chancellor of the University of California at Santa Cruz."

The assembled crowd of students, faculty, staff, other chancellors, and university officials burst into applause. It was a moment to savor—and to ponder. How had a boy raised in a lower-middle-class family in Chicago, whose forebears had never attended college, become the chancellor of the newest, most beautiful campus of the University of California? And how would he—this scientist and scholar, this somewhat reclusive academic—succeed as a chancellor?

"And, what college are you from?"

"Cowell College, but I live in town."

"I thought all freshmen lived in the college."

"They do, but I'm a senior."

"If you are a senior, why are you at the freshman reception?"

"Oh, I like to come every year to meet the new chancellor."

At Santa Cruz, the chancellor holds a reception each fall at University House for the new freshmen students. This was my first such occasion. It was a warm, clear evening and the students had spread out on the lawn before University House, munching on snacks, sipping soft drinks, chatting, and enjoying the moonlight reflected off Monterey Bay. I was drifting from group to group, making acquaintance and noting the enthusiasm and high spirits among the new students as they began a great adventure. (How quickly this wears off after classes begin and as the older students begin to transfer their cynicism and disillusion.)

The upperclassman—actually she was a fifth-year student (not many finish in four years at public universities)—set me back for a moment. She was nearly correct. In the fall of 1973, she would have met Dean McHenry, the founding chancellor, who resigned the next year. In 1974, it would have been Mark Christiansen, who resigned after eighteen months. In 1976, she would have met Angus Taylor, who served as interim chancellor. And here was I. Her remark brought home the administrative turmoil the campus had experienced in recent years and the powerful need for a period of stability. Yet the basic problems underlying the turmoil had to be resolved.

Soon after the fall quarter began, I became aware of a student rally that was being held in a small court near my office. The "cause," it soon appeared, was the denial of tenure, late the previous spring, to a popular assistant professor who was gay. The protestors were convinced that the denial was the consequence of his sexual orientation. This was a foretaste of how issue after issue would be soured and distorted by the claim of discrimination. The protestors presented me with a "demand" to reverse this decision. Although the case had been decided definitively by the previous chancellor, I agreed to review the file to assure myself the issue had been handled fairly. I did so and in the end fully agreed with the prior decision.

This announcement provoked another outburst, which ended with a sit-in of protestors in the chancellor's office. How should I deal with sit-ins, of which this would surely be only the first? Legally speaking, after 5 P.M. when the office closed, they were trespassing and I could have them arrested. Such action, however, elicited sympathy among other, less-involved students, hardened anti-administration feelings, and was practically ineffective. The previous chancellor had had several hundred students arrested the previous spring for a sit-in (against "institutional racism") over Memorial Day weekend. However, the local district attorney declined to prosecute any of them. The students had the right to demand individual trials. The community regarded the whole matter as student high jinks and was not about to spend hundreds of thousands of dollars on such trials, which would tie up the local courts for months. As I was repeatedly to learn, chancellors have little real power.

This sit-in was clearly a test of the new chancellor. I decided to wait the students out. As long as they did not actually interfere with the work of the office, they could symbolically sit in for as long as they pleased. I expected they would become bored, once the first enthusiasm waned; besides, Christmas vacation was coming. After I brought in some leftover goodies from our Christmas staff party at University House, the sit-in dissolved. Actually, they were nice kids just acting out.

I followed this policy of waiting out sit-ins throughout my tenure as chancellor. Although on one or two occasions it became quite tedious, ultimately it always worked and left better feelings and mutual respect. The important principle, it seemed to me, was the ability to carry on the work of the university, even if subject to minor inconvenience. On only one occasion did I ever have students arrested, when they blockaded the road entrance to the university. I had informed the group that

I could not allow them to bar other students, faculty, staff, or visitors from the campus. But they wanted to be arrested, and were. And, as could be expected, they were not prosecuted.

I had weekly meetings of the chancellor's immediate staff—vice-chancellors, personal assistants, a secretary, and other staff invited to discuss specific problems. At such a meeting, soon after I assumed office, the assistant chancellor for planning brought up a "very important" matter. What enrollment should we project for UC Santa Cruz for the fall of 1978? Having just begun the fall quarter of 1977, the topic seemed surprising. However, I learned that the president's office needed this number to prepare the proposed university budget for 1978–79, which would go to the board of regents in November for its approval. After that, it went to the governor's office and, one hoped, would be included without too much change in the budget he presented to the legislature in January 1978.

The "problem" was that any honest projection indicated that the expected enrollment in the fall of 1978 would be less than it was in the fall of 1977. It was then that I was acquainted with the deplorable slide in freshman applications to Santa Cruz that had begun in 1971 and continued each year since. By the use of varied tactics and alternatives, the campus enrollment had, however, actually been increased each year, until now. The "bag of tricks" was now empty. How much lower would the enrollment be in the fall of 1978? About one hundred students less than the current enrollment of fifty-seven hundred.

At Caltech, which had always rigidly limited its freshman class size, enrollment had never been an issue. Although I was strongly advised to the contrary, I found it hard to believe that an enrollment decline of less than 2 percent at Santa Cruz would be perceived as a serious problem. I was quite wrong. Indeed, this "problem" foreshadowed conceptual clashes between the tolerant flexibility of academia and the intolerant rigidities of the state accountants.

I opted for an honest estimate, which was included in the president's operating budget. The capital portion of the budget included some construction at Santa Cruz that had been put over from previous years and that was needed to adequately accommodate our present enrollment. Normally, on receipt of the proposed university capital budget, the governor's office and the office of the legislative analyst send a joint team to each campus to assess the need for the proposed construction and its scope relative to the need. These visits are usually rather challenging, and the campus is required to justify the request in detail. This year, however, the usual visit was abruptly cancelled. Clearly, they said, there was no need for new construction at a campus that was declining in enrollment. Period. The Santa Cruz construction was simply omitted from the governor's proposed state budget.

I had not been aware that the turmoil at the campus, combined with the marked fall-off from its early popularity, had given rise to rumors that the campus might well be closed by UC. Without the growth that had been predicted in the older demographic projections, the UC system did not really need eight general campuses at this time. The anticipated decline in enrollment for 1978 now revived these rumors in full force.

As might be expected, such an atmosphere made raising funds for the campus difficult. Indeed, such rumors have all the attributes of a self-fulfilling prophecy. Who wants to donate to a campus that might cease to exist? A specific consequence soon developed. The National Science Foundation wanted to establish an ongoing institute for the study of theoretical physics and was soliciting proposals from universities that would like to host this institute. The physicists at Santa Cruz were eager for this opportunity and the campus submitted a proposal. We identified space to house the proposed institute and agreed to provide some financial support and an academic position for the director. We talked with some distinguished physicists about their possible interest in becoming director should the institute come to Santa Cruz.

We were then disconcerted to learn that while many features of our proposal were favorably regarded at NSF, they had heard the rumors and were naturally concerned. They would not want to locate such an institute at a campus about to be closed. I contacted President Saxon, who reassured me that there was no substance to these rumors and agreed to write a letter to NSF to that effect. He did so, but we learned that NSF was not fully convinced. In the end, the institute went to a rival proposal from UC Santa Barbara. Other factors surely entered into this decision, but the taint our proposal suffered from this circumstance was undoubtedly fatal.

All of this convinced me that the enrollment problem had to be addressed promptly, and that to address it would require a major change in the image of the campus throughout the state. To improve the image, I had also to improve the substance. This required the cooperation of the faculty and even, in many cases, a reorientation of their personal goals.

The faculty hold a remarkable degree of authority in the UC system.

Control over the curriculum and educational policy and over admissions criteria are delegated by the board of regents to the faculty. While the administration—the chancellor—retains fiscal authority and ultimate promotion approval, it is accountable to the faculty for specific decisions. If there is not a meeting of minds, an agreement on a common set of values, the stage is set for continuing rancor and academic stalemate. In this pattern of "shared governance" the chancellor can propose but, except in rare instances, cannot implement without the active consent of the faculty.

While the chancellor has very limited real or immediate power, he does have a platform from which to advocate policies and standards. At the right moment, this can provide a critical leverage. And on the administrative side, he can over time, by well-chosen structural and personnel changes, strengthen and improve the implementation of his policies and standards to benefit the entire campus.

The interminable and continuing college-board conflict had not only sapped the energies of the faculty but, by confounding the criteria for performance, had pulled down the standards for faculty advancement and promotion. The most important advancement step is the award of tenure; once awarded, it can only be revoked by the board of regents for exceptional cause. Tenure had *never* been revoked in the 120-odd-year history of UC. (This is not to say that, in extreme cases, faculty members had not been persuaded to resign.) In the UC system, tenure must be awarded after eight years of service or reappointment denied, though it can be awarded earlier if merited. Since nonrenewal of appointment requires a year's prior notice, a decision as to tenure must be made before the end of the seventh year. This judgment requires an extensive review process, accompanied by a complex system of procedural safeguards to ensure fairness, absence of discrimination, opportunity for rebuttal, and so on.

As chancellor, I personally reviewed all tenure cases as they came up from the board of study (and, initially, also the college), and through the screen of the Academic Senate Committee on Academic Personnel (CAP). With frequently conflicting board and college recommendations and divided CAP recommendations, many cases required careful analysis to sort out the bases for each view. Most often, of course, tenure decisions had to be made in fields in which I had no particular expertise. I had available to me the file containing the candidate's record and work, external and internal evaluations, and the recommendations of the various levels of review. Starting with all of this background, I would still seek to form an independent opinion—especially if confronted with divided recommendations—by perusing the candidate's work directly. My background as a scientist and especially as an editor served me well. In most cases, I felt that I could analyze the logic—or the gaps and flaws therein—of the arguments presented in the candidate's work, evaluate the cogency and organization of the presentation, and ascertain if the candidate sought to relate his or her contribution to the precedents and larger issues in the field. (Of course, in some fields such as advanced mathematics and artistic creation, I was largely reduced to secondary judgments of the strengths and weaknesses reflected in the evaluations of others.) While clearly the Caltech criterion of national distinction would not be appropriate here, I felt that some clear evidence of competent scholarly performance with indication of continuing achievement—and not just "promise" or expectation—was a minimal requirement. This might be relaxed in the case of a truly exceptional teacher, but such cases could be expected to be rare.

This standard, which seemed reasonable if not generous to me and essential to the future of the campus, was clearly higher than that to which the campus had been accustomed. In my first year I overrode, negatively, four tenure recommendations from CAP and several other advancements. In each case, I documented my reasoning. These actions produced some consternation and grumbling, but they had the desired effect. In subsequent years, the standards of the CAP improved and the number of conflicting judgments diminished to almost nil.

I inherited a cadre of administrators—vice-chancellors, assistant chancellors, provosts of colleges, registrar—of varying ability and (worse) varying philosophy as to the mission of the campus. The college-board conflict even permeated the staff.

Because UC is such a large and distinctive system, it breeds its own form of professionalism. The administrative personnel at various levels, from the several campuses, periodically get together to discuss their common problems. Thus, the academic vice-chancellors meet, the financial planning vice-chancellors meet, the vice-chancellors for student affairs meet, the registrars meet, the student health directors meet, the financial aid officers meet, the affirmative action officers meet, and so on. While the exchange of information, ideas, and procedures is often valuable, these meetings tend to consolidate a sense of professional identity. I had to repeatedly point out to the administrative cadre that the business of the university was education and research, not administration, and that the administration was there to further the primary goals and not vice versa. This principle sounds so obvious, but to enforce it requires continuous attention.

My inherited staff was most helpful in initiating me in the often baroque ways of the UC system, but in some cases unfortunate individual divergences of working pattern and style became evident. A few could simply not alter the specific working patterns they had developed under my predecessors. Alternative positions were found for some; others chose to resign.

The vice-chancellor for academic affairs (AVC) is the second most important administrative position on the campus. The several deans report to the AVC and he or she must adjudicate among their many and varying requests in the allocation of always limited resources. The AVC also interacts closely with various senate committees. The AVC I inherited was a fine person and scholar but was uncomfortable with the difficult decisions the office entailed. I needed to find a replacement. This led to a painful situation.

The choice of AVC is made by the chancellor because the two must work closely together. The complexity and distinctive character of UC personnel and budgetary processes almost necessitated that the AVC have had prior UC experience. As I became acquainted with various campus personalities, I kept the AVC position in mind. By early spring, I had decided that the chair of the academic senate, Professor Paul N., would be a good fit. He was a straightforward, thoughtful person and a social scientist (I did not want both top officers to be natural scientists). By virtue of his position as chair of the senate, he was acquainted with almost all of the campus issues. I informally offered Paul the position and he accepted.

To my astonishment, a furor erupted. I was visited by several delegations of faculty. It developed that Paul's scholarship was poorly regarded on the campus. And, above all, it was felt that the AVC had to have high personal standards of scholarship. But Paul had been elected chair of the senate, I said. Oh well, they countered, the senate is unimportant. The AVC position is.

It became clear that Paul's appointment would provoke a violent and derogatory discussion in the senate, and that, lacking the respect of a major contingent of the faculty, he could never adequately fill the position. I had to ask him to withdraw his candidacy. It was a lesson to me of the need to consult more broadly and, as well, of the (lack of) repute of the academic senate.

Fortunately, Professor John Marcum, another candidate whom I had considered favorably, but who had told me he was unavailable because of other commitments, became available. His appointment was widely accepted.

Having steadily increased the student-to-faculty ratio from its initial low value to the systemwide value, the campus could no longer afford the dissipation of faculty effort in soft, "Mickey Mouse" types of courses taught only to fulfill a college commitment. Provision of these courses of varied merit was vitiating the disciplinary curricula. The basic problem—foreseen by some, but blandly ignored by others—of the greater cost of a true collegiate system was weighing more and more heavily on the campus. I took my first year to become fully acquainted with the dimensions of this dilemma and to explore, conceptually, various options. I had about resolved to "cut the Gordian knot" when external events at the end of that year forced the issue.

In June 1978, the voters of California passed Proposition 13, which abruptly more than halved the revenue available to city and county governments from property taxes. The state government then had immediately to provide the difference, lest all governmental agencies collapse. While the state government had been running a mild surplus, this was quite inadequate; the budgets for all the normal state functions, including higher education, had to be cut immediately to make up the funding.

This episode was a dramatic lesson for me of the ways in which the fortunes and plans of those in a public university are grossly affected by events wholly beyond their control—by the ephemeral passions of the voters, by the state of the economy, by the inclinations of the governor or the legislature. There truly are "windows of opportunity" when all of the conditions are propitious and the signals are "go," but there are also times when the windows are firmly closed. One must make plans and be ready to implement them during the periods of opportunity. And one must be prudent, alert for a sudden change in wind.

But now the windows were closed, and were to remain so for four years. Adaptation of the state budget to Proposition 13 required several years. Governor Jerry Brown was basically unsympathetic to the university. And the state's economy fell into a serious recession in 1981–82. The budgetary crisis of 1978–79 and the demoralizing decline in enrollment in the fall of 1978 forced me to propose in that fall the "campus reorganization" that I had conceived the previous spring.

Academic standards are the sine qua non of a major university. To maintain and improve the standards of the disciplines was a primary step in the recovery of the image of Santa Cruz. To accomplish this, the diversion of resources into second-rate college courses had to be stopped. The influence of the colleges had to be removed from personnel decisions so that the faculty had a clear set of academic goals. The vitiation of the intellectual life of the disciplines caused by the dispersion of faculty among all eight colleges had to be ended by establishing intellectually coherent groups of faculty in each discipline in at most two or three colleges.

At the same time, I wished to maintain the colleges as intellectual and cultural centers in the liberal arts tradition, as well as residence facilities. There would be several diverse groups of faculty in each college and a mix of students with varied interests. Each college would be required to provide a freshman "core" course on some broad topic of interest to its faculty—funds would be provided for this purpose. As available, funds might also be provided for other college-based endeavors that fell outside the scope of any discipline.

While I knew that this drastic proposal would provoke diehard resistance from those who would regard it as a betrayal of the Santa Cruz vision, the "dream," I believed that the majority of the faculty would welcome it. The proposal would provide a clear sense of direction and, finally, an end to the college-board wrangle. I hoped it would "jump start" the campus and release the energies of the faculty from its sterile internal strife. I checked that this would meet with the approval of the systemwide administration and then presented the proposal at a senate meeting just prior to my official inauguration in October 1978.

I had to have the senate's agreement. The debate continued through much of the academic year. I had appointed a faculty committee to work out the many details involved of faculty transfers between colleges, of course changes and curricular realignments, and so on. However, I found that I had to repeatedly take the initiative and propose solutions for seemingly thorny issues. In negotiating this thicket, plunging through inertia and sniping resistance, I had a psychological trump. My ego was not invested in the chancellorship. I was always prepared to resign and return to science if I could not lead the campus in a desirable direction. This sense of independence sustained me on several difficult occasions. In the end, when the academic senate voted in the spring, over 80 percent were in favor.

There was really no alternative. But it was, in many ways, the most trying and exhausting period I have known. It was a period of crisis for the institution, which required leadership, determination, a clear vision, and at least the appearance of patience. I succeeded, but there was a small residue of faculty who were never reconciled. Despite the evidence, they could not accept that they had devoted years of effort to an impractical vision. That summer we moved the offices of more than half of the faculty in the colleges. By fall we were on the new path.

This tale illustrates the critical importance of a sound beginning, especially in academia, where experimentation is foreign to much of the faculty and the diffusion of authority defies change. In this light, the casual failure to have thought through the basic initial conception of Santa Cruz is all the more deplorable.

The retrenchments required by the budgetary stringency over the next several years were painful, but their effects were not all bad. They provided immediate justification for organizational changes that achieved greater efficiency and economy, and their pervasive nature brought administration and faculty into closer consultation than had been the custom, a practice that was subsequently continued.

The enrollment problem had to be addressed and soon. I recall dreary afternoons that fall spent in discussions of which dormitories we might have to close the next spring. And campus images in academia are not changed abruptly. While "recruiting" of students by UC campuses is officially frowned on, "outreach programs" to provide prospective students with information about the opportunities available on a campus are sanctioned and practiced. Regrettably, because Santa Cruz had been so popular in its early days, the campus had not bothered to develop an effective outreach staff and program. And during the years of decline in the early and mid-1970s, the campus had been too preoccupied with its internal crises.

I needed an outstanding dean of admissions and happily found Richard Moll. A Yale alumnus, Dick had been dean of admissions at Bowdoin where he reversed their enrollment decline. Then he managed the admissions program at Vassar during their complex transition from women's college to coeducational college. Dick was a vibrant, gregarious person fond of playing the piano and warbling old favorites. One of the most skilled and adroit admissions directors in the country, he was looking for a new challenge. Having worked only in private colleges, the problems of a distinctive, relatively small public university— a "public Ivy," as he called it—intrigued him.

Moll immediately invigorated our outreach program. With tasteful brochures and a comprehensive program of visits to likely high schools, public and private, throughout the state, through involvement of alumni, he spread the word about the potential and quality of the campus with its new directions. To change a public image takes time and requires some real accomplishments, but over the next three to five years, Dick's efforts paid off. By 1983, enrollment applications were rising and have accelerated ever since.

In part, these increases reflected the growing high school population in California. But when Santa Cruz applications exceeded those at Davis, we knew that Dick Moll's efforts had succeeded. Moll also mentored Joe Allen and groomed him to be his successor. His primary job done, Dick went on, but his legacy remained.

A Chinese blessing says, "May you live in interesting times." These were certainly "interesting" times and full of action, but often they were also times for introspection. The chancellor, I had soon realized, is the one person who is charged with the welfare of the entire campus. Everyone else has a more parochial interest. And at Santa Cruz, I had soon relearned the old lesson that in human affairs logic is not enough. Simple logic does not overcome passion, prejudice, or fear. One can use logic to divert passion, to play fear against fear, or even invert prejudice—in short, to play politics—but this is devious logic that does not come naturally to a scientist.

In contrast to the intellectually exciting but academically serene ambience of Caltech, I suddenly found myself in a world of student protests, enrollment and fiscal crises, and system politics. I was now at the head of a demoralized campus founded (and foundering) on an ill-conceived plan, woefully maladapted to the ethos of the larger university. My original academic goals were stymied behind an array of immediate problems, and especially lack of resources and lack of true educational authority.

I felt confident that I could redirect and revive this campus and restore its inherent potential, but it would be a task of many years and great effort with steadfast purpose. My MIT education, my scientist training, would be my resource. To be a problem solver requires an objectivity, a skepticism of preconceived ideological answers, and a committed integrity that eschews self-deception that clearly recognizes failure as well as success, that invents or improvises, bends or tacks, but always knows the goal.

I kept at it. Why? Was it stubbornness, an inability to admit a career error? I think not. Scientists understand failed experiments. Was it an altruistic impulse to use my talents to somehow save this campus? Was it a recognition of my being fifty-seven years old and wanting to make the best use of the next ten years? Was it simply the challenge to beat the odds, to do the nearly impossible, to leave an enduring achievement, to create something different and special? Possibly all of these and more. In some deep sense, what I was doing, while not always enjoyable, was right—for myself and for Santa Cruz.

22

The Club of Nine

The chancellors of the nine UC campuses are a diverse and uncommon lot. An exclusive club, all male in my time, they are bonded by their common problems and by a common concern (genuine but not disinterested) for the welfare of the university. Yet they are necessarily separated by their inherent loyalties and obligations to their individual, competing institutions. All able, selected with care, they have diverse talents, personalities, and experiences. With many tasks in common, they also face dissimilar circumstances and opportunities, with varied degrees of success. Collectively, they illuminate the virtues and flaws of the finest system of public higher education in the nation.

The two oldest and largest campuses, Berkeley and Los Angeles, dominate the system. Together, for most of my tenure as chancellor, they comprised roughly half of the UC system and well over half of its resources. Their chancellors tended to dominate our meetings. It soon became evident that, while they could not always obtain what they wanted, nothing could be adopted over their opposition. While, of course, nominally subordinate to the president, the chancellors of Berkeley and Los Angeles had their own constituencies—large alumni bodies, large faculties, major sources of gift funds, direct relations with the many regents in the Bay Area and Los Angeles respectively. The president, while definitely not their captive, would not readily or often pursue actions directly counter to the desires of the Berkeley or Los Angeles chancellors.

When I became chancellor at Santa Cruz, Albert Bowker, a statistician, was the chancellor at Berkeley. Earlier, he had been dean of graduate studies at Stanford and then chancellor at the City College of New York. A quiet, bulky man, he often sat Buddha-like at chancellors' meetings "playing his cards close to his chest," presenting quietly but firmly his opinion after most of the others had spoken. He defended the Berkeley interests well. Berkeley had for many years been *the* University of California and it still retains that self-perception.

In 1980, Bowker resigned to become assistant secretary for postsecondary education in the Department of Education in Washington. He was succeeded by Ira Michael Heyman, a law professor who had been provost under Bowker. The search for Bowker's successor had been flawed by failures of confidentiality and it was well known that the leading candidate had withdrawn at the last minute because of "leaks." Michael thus came into office with this slight handicap, but it did not hold him back. A tall, sturdy man of liberal persuasions, Heyman was well acquainted with the problems of Berkeley and set about to remedy them. Perceiving biology as the field in which Berkeley should establish its preeminence in the succeeding decades, as it had in physics in the mid-twentieth century, he undertook a complete campus reorganization of the fractured biology program and diverted resources in that direction.

As the most popular of the UC campuses, Berkeley was overwhelmed with freshman applicants. At the same time, because of attrition in the freshman and sophomore years, it had excess capacity at the junior and senior levels. In the years of underenrollment at Santa Cruz, I made an arrangement with Heyman that Berkeley would refer a number of freshman applicants, who were eligible for admission but whom Berkeley could not accept, to Santa Cruz with the commitment that if they did well at Santa Cruz for two years they could then transfer to Berkeley with automatic acceptance. For five years, this program sent about two hundred freshmen to Santa Cruz. Interestingly, after the two years at Santa Cruz, about half chose to remain.

Chuck Young, the perennial chancellor at Los Angeles (in his twenty-fifth year at this writing) is also quite tall and vigorous, a former football player at Riverside. Never a real scholar, Chuck is gregarious, a superb fund-raiser with the community and alumni, an enthusiastic promoter of athletics, and a good manager. Wisely, he has appointed able vice-chancellors who have handled most of the academic affairs. Tapping the immense resources of Los Angeles and especially the wealth of adjacent communities such as Beverly Hills, Bel Air, and Brentwood, UCLA has grown mightily under Young's stewardship. Ever resentful of Berkeley's primacy and scholarly eminence in the UC system, UCLA has made little secret of its desire for relative independence from the strictures of the system. If it could somehow be given its state funding unfettered, UCLA would like simply to go its own way—and it could.

By virtue of the stature of their campuses and their own physical stature, Heyman and Young invariably played major roles at the meetings of the council of chancellors.

The other chancellors each had distinctive personalities. Chancellors tend to emphasize one or a few facets of their multifaceted task. Academic leadership and planning, campus management, fund-raising, UC system (and campus) policy determination and politics, community and alumni relations, participation in the national scene within academia or within their professions—all are appropriate and necessary functions. Priorities may vary in relation to the maturity of the campus and the opportunities and importunities of the time, and each chancellor is inclined to favor one or another depending on his or her particular interests and talents.

In one important respect, scientists may not make good chancellors. They can be objective and they have the integrity essential to cope with an unfeeling nature. But as problem solvers they tend naturally to focus on the *substance* of an issue, to devise and analyze various options for dealing with it, and to decide on the most effective. However, substance is often not the heart of the issue. The issue may be primarily symbolic or, to the participants, procedure may be more important than outcome. Both concepts are foreign to the scientist, who deals with concrete, not symbolic problems, and for whom niceties of procedure are unimportant—"whatever works!" In this realm the skills of the lawyer and the politician are often the most relevant. Scientists can, of course, learn to think in political and symbolic terms; once properly formulated, the skills of problem solving are quite applicable in these domains. But scientists may well find such arenas and tactics uncongenial.

Jim Meyer, the chancellor at Davis, exemplified emphasis on one facet—management. With limited interest in academic questions, he delighted in the details of management structure and function, seeking economy and effectiveness. In an organization as large and bureaucratic as UC, such talents were valuable, but they did not arouse great enthusiasm on his campus.

Francis Sooy and his successor Julius Krevans were the chancellors at UC San Francisco during my tenure. UC San Francisco is a specialized campus focusing on medicine and associated fields. Both Sooy and Krevans were distinguished physician-researchers skilled in the politics within the medical profession. Both had high standards and should be given much credit for the development of that campus into a research and teaching institution at the very top level of U.S. medicine.

Bob Huttenback, the chancellor at Santa Barbara, was a longtime colleague. An historian, he also had been a division chairman (of humanities and social science) at Caltech and went to Santa Barbara shortly after I went to Santa Cruz. Huttenback was bright, overconfident, and fond of good food, with a strong personality and a streak of vanity. He had high standards, and at Santa Barbara he had to implement some difficult renewals. A former teachers' college that had earlier been converted into a UC campus, Santa Barbara had had areas of difficulty in its emergence from its less distinguished origins. Bob attacked these problems forcefully, bruising egos in the process. He set about to improve particularly the physical sciences and engineering and boosted these into national stature. However, he also tended toward grandiose projects of marginal academic importance such as a joint ecological project with the city government of Venice, Italy, and a projected Food and Wine Institute on the campus.

The wives of chancellors can play an important if unheralded role in their success or failure. Huttenback's wife, Freda, played a minimal role on campus. At her insistence, the Huttenbacks moved out of the university-provided chancellor's residence on campus to a private house in an elegant part of Santa Barbara. This latter proved to be Huttenback's undoing. Over the years, over two hundred thousand dollars of university funds (above and beyond a housing allowance) were spent to renovate, maintain, and adorn the Huttenbacks' private residence. When this became known, his enemies, who had been awaiting an opportunity, struck. He was obliged to resign, brought to trial for "embezzlement" of public funds, and convicted.

This was a tragedy of folly. I am convinced Huttenback had no criminal intent and had someone pointed out that this use of funds could be construed as "embezzlement," he would have been stunned. Folly, yes; criminality, no. And a sad outcome, for in truth he did many good deeds for UC Santa Barbara and surely improved its status and prestige.

The chancellorship is a very public position. I came to realize this early on from casual comments about my personal actions. And the chancellor, as the one who makes the really difficult decisions, is bound to create unhappiness, even enmity. For this reason I consistently "leaned over backward" to avoid any interpretation of impropriety. The chancellor's role inevitably engenders ambiguities and gray areas of proper allocation, but if a question arose as to whether an expenditure was personal or institutional, I simply accepted the former. To do otherwise would have been pointless folly.

Dan Aldrich at Irvine, an agricultural scientist, was the paragon of a product of the UC system. After service as researcher, professor, and administrator at Riverside, Davis, and Berkeley, he became the founding chancellor at UC Irvine, which he led for twenty-two years. Tall, genial, and politically shrewd, very experienced yet not especially imaginative, he had the simplistic if grand ambition to make Irvine a clone of Berkeley. A physical fitness buff, he competed successfully in senior track and field contests into his seventies.

After his retirement in 1985, he became the UC "designated hitter," serving as acting chancellor for a year at Riverside and then again for a year at Santa Barbara. Both interim terms were successful. His technique for handling student protest in that time was simple and ingenious. "I agree with you and I will recommend so to the next chancellor!" Aldrich was succeeded at Irvine by Jack Peltason, a political scientist and experienced academic administrator, formerly chancellor at the Urbana campus of the University of Illinois and then president of the American Council on Education. Peltason particularly and successfully cultivated the wealthy community near Irvine in Orange County and raised large sums for the campus. Subsequently, in 1992, Peltason, experienced and noncontroversial, was selected to succeed David Gardner as president of the university.

Bill McElroy, a large, bluff, hearty, quick-tempered Irishman and skilled biochemist, was chancellor at San Diego. With its emphasis on science and technology, San Diego has frequently had chancellors with scientific backgrounds. Early in my chancellorial tenure, I served on a small committee, appointed by President Saxon, with McElroy as chair. Our mission was to review the "organized research" programs of the UC system and make recommendations for their improvement. This proved to be a lesson in UC politics.

UC is the "research arm" of California and the state provides funding to support various research programs. While the base level of this funding has not been increased in many years, it has been augmented annually to compensate for inflation and currently exceeds one hundred million dollars per year. About half of this sum goes to agricultural research at Davis and Riverside, while the other half goes to a diverse

set of "organized research" projects. Nominally, these projects are established to operate multidisciplinary research programs that extend
beyond the appropriate purview of a single department or that involve
extensive off-campus facilities or interactions. A marine laboratory, a
high-energy physics program, and the multicampus UC astronomy program are viable examples. The state funding is intended to provide stable core support while the bulk of the research is supported by external
grants.

Because most of the original funding for such organized research
units (ORUs) became available in the 1950s and 1960s, they were concentrated at Berkeley and Los Angeles. Once established, unlike programs supported by external grants, they had no termination dates, no
requirements for review, and no pattern of periodic competition for
these funds. Our committee found that many of the programs, which
had initially flourished, had become stagnant or outdated and of low
productivity. We recommended that ORU grants should be for an initial five-year period. After review, they could be renewed for a second
five years, but only rarely longer. We recommended that funds made
available by termination of an ORU—or alternatively by a small annual
decrement in the funding to all ongoing ORUs—be used for a competitive grant program throughout the UC system. We believed that
such actions would markedly improve the quality of research supported
with these funds.

Of course, the net effect of our proposal could only be to withdraw
and redistribute some funding from Berkeley and Los Angeles to other
campuses on the basis of competition. As a result, the chancellors of
Berkeley and Los Angeles were opposed and the proposal died quietly.
It was a lesson learned but never fully accepted.

McElroy's personality clashed repeatedly with that of President
Saxon, who was far more cautious and temperate. McElroy's somewhat
heavy-handed style created enemies at San Diego even within his own
administration, and when he tried to increase direct campus administrative control over the UC San Diego Medical School, its opposition
led to a vote of no-confidence. When Saxon did not support him, Bill
had no choice but to resign. He was succeeded by Dick Atkinson, a
psychologist who had been director of the National Science Foundation. Much more tactful than McElroy, Atkinson has maintained harmony on the campus while continuing to play a national role, serving
a term as president of the American Association for the Advancement
of Science and in the National Academy of Sciences. Good at fund-

raising, he has also been remarkably successful at using the growing population of the San Diego region and the presence of a regent from San Diego to achieve major resource and capital allocations to his campus.

The Riverside campus has long been the least popular in the system. It was initially established in the 1950s as a unique four-year campus within the UC system, grafted onto an agricultural experiment station. Later, it recognized its foreign character within the UC system and opted to become a full-scale university. But, located inland in a region with heavy smog, it was largely shunned by students. Ivan Hinderaker, a political scientist and another longtime UC staff member, was its chancellor until his retirement in 1983. He was succeeded by Tomas Rivera, a literature scholar and poet, who was the first person of ethnic minority ever appointed to a UC chancellorship. As Riverside is in a region of rapidly growing Hispanic population, his choice seemed felicitous. He was just taking hold of the campus when, tragically, a sudden heart attack took his life.

And what of the chancellor at Santa Cruz in this all-male group of mostly experienced administrators? Intellectually, I could easily hold my own, but this was not an arena in which intellect was always the primary asset. I had much to learn, and fast. Unfamiliar with the idiosyncrasies and bureaucratic complexity of the UC system, coming from a small, private, specialized institution into this vast public enterprise, I was initially taken aback by its unfamiliarity and, as well, by the depth of the inherited problems at the Santa Cruz campus. As a molecular biologist, I was able to bring a dispassionate objectivity to this task, which became recognized as integrity or "fairness." Trained from childhood to be wary of anger, to suppress it and allow it to drain away, I never learned to *use* anger as is sometimes appropriate.

I was never fully comfortable with fund-raising, even for fully deserving causes, and did it reluctantly. While recognizing the sometimes overriding importance of political maneuver, I found it uncongenial and engaged in such tactics sparingly. Preferring to make my case with logic, I sometimes failed to achieve all that I might have for my campus.

Perceiving the need, I concerned myself greatly with the academic programs and standards of the campus and there made my major mark. After a life spent in the orderly, single-valued world of basic science, I found the more complex worlds and motivations of the other academic divisions—the social sciences, the humanities, and the arts—both interesting and odd. The several academic divisions had varied perspectives and values, even differing concepts of "truth." The significance of time varied greatly across the campus. The rapid progress of the sciences necessitated frequent changes of course material, curricula, and research emphases. Many of the social scientists similarly needed to at least keep pace with the tempo of societal change. But in the areas of philosophy, political science, and parts of literature, scholars are still grappling with the same basic problems that troubled Plato and Socrates. And artists, too, although sometimes with new media, conjure with ancient issues of form and expression.

As scientist-chancellor, I tried to meet and accept all of these diverse universes on their own terms but often found it hard to abandon my scientist's faith in a single objective truth, capable of withstanding challenge, each nugget of understanding in itself an increment of human knowledge.

The plight of the campus demanded my full energies. I found it impossible to seriously pursue my science. I gradually and reluctantly withdrew from activities on the national scene, which did not seem of much benefit to my campus and, instead, seemed to distract my attention and effort. The campus, so badly fractured, sorely needed a steady and guiding presence. In dark hours, when the target of the perennial revolt of youth against authority or the object of unhappy regard by this or that disappointed segment of the faculty, I wondered at the wisdom of my choice. But my commitment had been given and to withdraw it could only be deeply demoralizing to this already betrayed campus. And I learned early not to take the attacks personally, for most often it was the chancellorship itself, and not the holder of the office, that really was the target.

The chancellors who personally thrive and most enjoy the role, who find it most rewarding on a day-to-day basis, are those who like to persuade, to match wits with and best their adversaries, to manipulate others, to use the adroit assessment of personality and choice of incentives or discouragements to achieve a desired end. In short, they are those with the inclinations, natural or learned, of the salesman or the politician. Those whose minds and energies are seized more by the issues, whether abstract or substantive, find the role often drearily frustrating and eventually wearing.

23

Two Presidents

"The problem with you scientists is that you do not realize that the answer is not necessarily the solution." How true and how astute. I was in a conversation with David Gardner, the president of the University of California. In the "real world," the world in which he must function, the *answer*, the most efficient outcome of an analysis, might well not be the *solution* because it is simply politically infeasible. The solution must be an answer that is attainable, and Gardner is a master at finding such solutions.

And, I think, his "you scientists" included his predecessor, David Saxon, a physicist.

I served under two presidents of the UC system—both highly intelligent, dedicated, complex persons with very different backgrounds, experiences, styles, and modes of thought. The presidency of the UC system is a task of extraordinary complexity. Located at the nexus of the often conflicting expectations of the regents, the faculty, the students, the alumni, the staff, the governor, the legislature, and the general public (including many special interest groups), the president is also the "manager" of a ten billion dollar per year enterprise. The president is the one person whose mandate is the welfare of the university as a whole, beyond the aggregate of its components.

Presidents have their individual predilections and values. They may have a warm spot in their heart for one or another of the campuses. They may value some aspects of the university's intellectual endeavors more than others. They may feel more obligation to administrators, such as chancellors, whom they have chosen, than to those they have inherited. But the extreme diversity of the constituencies to which they are responsible constrains much deviation from accustomed patterns, and thereby limits innovation.

I was appointed by David Saxon. As has been mentioned, Saxon was a classmate at MIT. After graduation, he went on to receive his Ph.D. in theoretical physics from MIT. In 1947, he joined the faculty at UCLA. Although never a distinguished physicist, he authored several successful textbooks, including a *Physics for Liberal Arts Students*. Over the years, he displayed a talent for administration, becoming successively department chairman, dean of physical sciences, executive vice-chancellor at UCLA, and then provost of the university for the entire system. In 1975, he was chosen by the board of regents to become the fourteenth president of the University of California.

Saxon brought a scientific and academic perspective to the presidency. This was a distinct change from the style of the previous president, Charles Hitch, an economist deeply involved with fiscal concerns, cost-benefit analysis, and so on. As a scientist like myself, Saxon's attention focused first on the substance of an issue. While he recognized that the protagonists for a position might not hold very deep (much less reasoned) convictions, he was not likely to seek a mollifying symbolic or public-relations resolution. Instead, he often saw a current issue as a symptom, a manifestation, of a deeper problem than the one at hand and attempted to address that. Unfortunately, this more basic approach often did not satisfy the activists, who felt bypassed and ignored. Such reactions understandably left Saxon frustrated. While cognizant of political considerations, he did not weigh them heavily in his analysis, preferring to believe that logic would carry the day. He was not afraid to enter the political arena, however—on a logical platform, of course.

In 1980, following the success of the Proposition 13 initiative of 1978, the same political forces, seeking to contract the role of government, put forth an initiative to halve the state income tax. The effect on the state budget—and hence the university—already staggering under the impact of Proposition 13, would have been devastating. While many groups around the state could foresee and deplore the potential impact of this initiative, the overwhelming vote in 1978 for Proposition 13 seemingly paralyzed all potential opposition to the new proposal. Saxon took the initiative to organize a coalition to oppose this measure. He rounded up support from the business community, the labor forces, the various groups of state, county, and city employees, and other good government advocates and mounted a vigorous educational campaign that, in the end, defeated the proposal. His leadership, coming from an unlikely source, was crucial in this effort.

As a scientist, Saxon could immediately grasp the significance of a basic research program such as the project to build a ten-meter telescope (see chap. 29). He was enthused by the idea and saw its accomplishment as a major achievement of his administration. He supported this endeavor through its early stages, even though the university budget was in a grievous state. He believed it was always important to invest in the future and cited the example of UC President Robert Gordon Sproul, who in the midst of the great depression of the 1930s found one million dollars to support the construction of E. O. Lawrence's early cyclotron. This investment secured the preeminence of the Berkeley physics program for three decades.

It was Saxon's misfortune to be president during the governorship of Jerry Brown. A pure political opportunist with a vaguely utopian philosophy born in the 1960s, Brown maintained few consistent positions, one of which was an antipathy to the university, which he regarded as stodgy, elitist, and arrogant. He pinched the university budget even while the state was running a considerable budget surplus and after Proposition 13 seriously eroded university funding.

Brown also changed the composition of the board of regents, appointing several members of much more politically liberal tendencies. He also, in the guise of diversifying the board, made several highly idiosyncratic appointments, including an iconoclastic, retired UC philosophy professor, the widow of a distinguished UC anthropologist, and the director of a popular music recording company. The new appointees differed, often markedly, in ideology and personality from the older, more "establishment-type" regents, resulting in frequent clashes over policy and even procedure. As president of the university, Saxon was both a regent and the prime employee of the regents. In his unique position, he was expected to mediate the regents' internal conflicts and find effective compromise positions. This unsolicited task at times exceeded his political skills. The result eroded regental confidence in him and their own self-esteem. At the same time, the serious consequences of the budget cuts exacerbated internal competitions and rivalries within the university, leading to dissension and wrangling.

Feeling increasingly besieged and unappreciated in an increasingly charged political atmosphere, Saxon elected to resign in 1983. He went on to a far more congenial role as chairman of the corporation at MIT, which he filled quite successfully for seven years.

An academic, Saxon never fully recognized the need for the leading university official to maintain a presidential style and ambience among leaders of the business, banking, and allied communities. In a fiscal sense, he "wore a hairshirt," which he rationalized by the poor fiscal condition of the university. He kept his own salary low and thereby— since he was the president—those of administrators throughout the university. However idealistic such an attitude may be, it sooner or later will trip over market forces. Thus, Saxon's successor had to be offered a salary nearly twice that of Saxon merely to match what he had been receiving at his previous post. This "extraordinary increase," of course, elicited many disparaging comments and was an initial and quite unwarranted embarrassment for the new president.

Saxon's successor was a very different type. David Gardner is a master political strategist. While a thoughtful person with strong and clear convictions as to the importance of the university and its particular and special role in society, his approach to issues involves the political component from the beginning and weighs it heavily. Gardner's career has been in academic administration. He has a Ph.D. in education from Berkeley. Over the years, he rose rapidly from director of the Berkeley alumni foundation to assistant chancellor at UC Santa Barbara, to vice-chancellor at UC Santa Barbara, to vice-president of the UC system, to president of the University of Utah. He chaired the National Commission on Excellence in Education, which produced its famous report *America at Risk* in 1983, and in the same year was chosen to succeed Saxon.

Both presidents solicited advice from their staff and from the chancellors before coming to a policy decision. However, as an academic Saxon understood if a chancellor disagreed and maintained an independent position so long as it was not one of overt opposition. Gardner, to the contrary, expected that once a policy decision was made, all chancellors would support it.

Perhaps because he had been at UCLA during its early years of growth, Saxon felt a major responsibility to support and assist the smaller campuses of the system, to the extent that he could, even at some political hazard. Gardner, ever more attentive to the political cost, was less inclined to do so. Within the system, he recognized the political clout of Berkeley and Los Angeles and, to an increasing degree, San Diego. The other campuses are not neglected but neither do they merit special resource, especially political resource.

As an academic, and particularly a scientist, Saxon could more readily recognize and correct a "mistake." After all, not all experiments succeed. As a professional administrator, Gardner had great difficulty acknowledging a possible error.

By personal contact and cultivation, Gardner calmed the feuds and angers within the board of regents. He developed good contacts with the legislature. A Mormon, he showed a remarkable ability to maintain his equanimity when Willie Brown, the African American speaker of the state assembly, sought to provoke him with stinging remarks about the attitudes of the Mormon church toward African Americans.

The political balancing act at which Gardner excels requires a carefully crafted script on important occasions. Consequently, he does not like "surprises," new results, or unforeseen objections that might snarl his scenario, especially surprises from his own "team." As part of his "script," Gardner has an exceptional ability to summarize, in a fair and coherent manner, the essential points of what may have been a lengthy and tendentious discussion. He then presents his position, taking into account the salient arguments on both sides. This bravura performance leaves each side satisfied that, at least, its position has been heard and understood, even if it loses the subsequent vote.

It was Gardner's good fortune to become president as a new governor, George Deukmejian, entered office. The two established a special rapport. Deukmejian, the son of Armenian immigrants, had always appreciated the opportunity he had had for higher education and strongly backed the university. Gardner persuaded him of the sore need to redress the damage Brown had done and to set a goal of leaving a strong university as a significant part of his gubernatorial legacy. Gardner plays political chess to achieve his aims. He husbands his political capital, using it sparingly and wisely. When an external issue of significance to the university arises, such as the legislatively mandated review of the California Master Plan for Higher Education, or the attempt by the California State University system to share with UC the prerogative of awarding doctoral degrees, Gardner carefully thinks through a political campaign—positions, ploys, tactics, personalities—to ensure the maintenance of the university's status.

For many years, the UC central administration was headquartered across Oxford Street from the Berkeley campus. This propinquity made the president's office a handy target for student protests over varied issues. The sympathy, if not outright support, of the Berkeley courts for student protestors made "defense" of the president's office difficult, even defense from physical assault. Also, the juxtaposition of the president's office and the Berkeley campus gave rise to confusion and problems of protocol when greeting distinguished visitors, especially from abroad. Was the French president François-Maurice Mitterand visiting the University of California or UC Berkeley? Should the president or the chancellor be responsible for his itinerary and entertainment? Saxon had simply endured these problems. Gardner, determined to emphasize the primacy of the UC system, moved his headquarters to nearby Oakland.

Of course, even Gardner's skills could not resolve all issues. The problem of the disproportionate representation of minorities in higher education is most acute in California. The rapidly growing Hispanic and Asian American populations together with the African Americans will make the combined "minorities" the majority in California within a decade or two. This will be true of graduating high school students even sooner. As a public university, UC should reasonably reflect the population of the state. Yet "Anglos" are still the clear majority (54 percent) of the student body and even more so of the faculty (85 percent), although women and minorities now comprise 51 percent of the junior faculty, so their proportions will continue to increase.

Under the master plan for higher education in California, the university is expected to set its admissions criteria so that one eighth (12.5 percent) of the state's high school graduates will be eligible to attend. A formula combining SAT scores and high school grade point averages has been devised that, historically, has limited eligibility to about 13 to 14 percent of the high school graduates. But while approximately 14 percent of "Anglo" graduates are eligible, some 32 percent of Asian American graduates are eligible as compared with only 5 to 6 percent of Hispanic American graduates and 3 to 4 percent of African American graduates. To partially compensate for such disparities, the university has relaxed its requirements to admit graduates who fail to meet the standard requirements up to a maximum of 6 percent of the freshman class. While this exception was originally introduced to permit entry of athletes, it is now used primarily for minority students.

At heavily over-subscribed campuses such as Berkeley and Los Angeles, a strictly meritocratic admissions policy would exclude almost all Hispanic and African American students. To avert this politically disastrous outcome and in order to provide a diversity in the student body, de facto quotas have been established for these groups. These "quotas" have been the target of ceaseless protest, directed especially at Gardner, by the Asian American community. Appreciable numbers of Asian American students in the Bay Area with outstanding academic records have been unable to attend Berkeley. The Asians claim, with justification, that they are being discriminated against to remedy past discrimination against African and Hispanic Americans in which they had no part.

While the university has "ducked and dodged" this issue, the fact is that it has for political reasons deviated significantly from purely meritocratic policies. Whatever the justification, in the end someone bears the burden of such discrimination. And someone, namely the university president, feels the heat.

More heuristically, the university, beginning under Saxon and continuing ever more actively under Gardner, has mounted a series of programs designed to increase the proportion of minorities, especially Hispanic and African Americans, in the university and ultimately on the faculty. "Early Outreach" programs, sponsored by the university, provide counseling and guidance beginning in junior high schools with predominantly minority enrollment. Manifestly, if students fail to take the appropriate courses in high school, they will not be eligible to attend the university. Special summer "bridge" programs help to acclimate minority high school graduates to the university practices and milieu. Special counseling and tutorial programs, primarily for minority students, are available on every campus. Special financial aid programs exist for minority students.

Recognizing that unless minority students attend graduate school there will be no pool of future minority faculty, the university has provided for special financial support for minority graduate students and even a unique program of postdoctoral fellowships for able minority and women Ph.D.'s. And the university has a well-conceived and implemented, although necessarily officious, affirmative action program to ensure that minority and women candidates are considered for all faculty and staff positions.

Nonetheless, with the long training period and the slow turnover, many years will inevitably be required for the representation of minorities and women on the faculty to reflect their proportions in the student body and the general population. In the meantime, the corresponding interest groups will direct their unhappiness toward the university administration and especially the president.

In contrasting these two presidents, there can be little question that, objectively, Gardner was the more successful. In that position, political skills are more significant than sophisticated analytical intelligence. And Gardner, the experienced professional administrator, was clearly more comfortable in the position, more self-confident and at ease, than the academic Saxon, who was more self-critical, more burdened by consciousness of the merit of opposing views. But how would Gardner have fared with Jerry Brown as governor? It would have been interesting.

In 1992, sorely grieving the sudden and tragically early death of his wife Libby, David Gardner resigned as president of the University of California. His departure was unhappily tarnished by the revelation of a generous retirement "package" provided to him by the regents, but awkwardly coincident with major recession-induced crises for the state and the university. In the short term, in a time of salary freezes, student fee increases, layoffs, and so on, this "golden parachute," however well merited, overshadowed his remarkable accomplishments throughout his nine-year tenure and sullied the image of the UC administration in the eyes of the legislature and the citizenry in general.

24

The Regents

It was a dramatic confrontation. Governor George Deukmejian, a candidate for reelection, sat as an ex officio member with the board of regents, next to the chairman. His election opponent, Tom Bradley, the African American mayor of Los Angeles whom he had narrowly defeated four years previous, stood before the board as a petitioner.

The issue was divestment. The question of continuing investment of university funds—endowment and retirement accumulations—in companies doing business in South Africa had bedeviled the board of regents for years. Student protests against such investment had erupted spasmodically, reaching a crescendo the past spring when several thousand had surrounded the regents' meeting at the Lawrence Exploratorium at UC Berkeley, requiring a protective police cordon and, after the meeting, surreptitious exit by a devious route.

The investment committee of the board, dominated by senior, conservative regents from the worlds of banking and commerce, had been loath to divest. Mindful of its "fiduciary" responsibility to manage these funds wisely and prudently, the committee was unsure that alternative investments of comparable quality could be found. It questioned whether the exodus of U.S. corporations from South Africa would benefit or harm the native population there. It also hesitated to open what some saw as a Pandora's box; once one began to discriminate among investments on the basis of perceived "social value," where did the process stop? What about companies that did business with the Soviet Union? Liquor companies? Tobacco companies?

President Gardner had supported this position and divestment had repeatedly been discussed, debated, and defeated. The regents had "compromised" by agreeing to urge all companies in which they held equity to subscribe to the Sullivan Principles, which encouraged American companies in South Africa to hire, train, pay equal wages to, and advance native workers as rapidly as possible. Personally, I favored divestment. But under Gardner's concept of chancellorial responsibility, I could only indicate my view quietly, eschewing public statement.

Mayor Bradley, promoting his election campaign, was now "demanding" that the regents divest. The governor, however, was ready with a preemptive strike. He had a day earlier announced that he intended to introduce a motion for gradual divestiture allowing corporations a period of grace to withdraw from South African enterprise before UC would be required to sell its equities. And this time the governor called in his political chips—all regents appointed by Deukmejian voted for the motion and it passed.

The board of regents is supposed to insulate the university from political conflict, but it is not always able to do so. It was fitting that this regental action, which evoked student joy, occurred at the first regents' meeting at Santa Cruz in eighteen years. After a near riot by students furious with Governor Reagan following a regents' meeting at Santa Cruz in 1968, the regents had ceased to meet on campuses for sixteen years. They had now somewhat gingerly resumed that practice for a few meetings each year and were once again at Santa Cruz.

The eminence of the University of California owes much to the wise traditions and relative political independence of its board of regents. By provision of the state constitution, the governance of the university is delegated specifically to the board of regents. Therefore, UC is administratively independent of the whims and passions of the state legislature. This is not true of the California State University system nor the community college system, both of which are creations of the legislature. The independence of UC, of course, has its limits. The legislature has the power of the purse through its appropriations and could therefore exert considerable pressure to achieve certain objectives. Fortunately, however, over the decades, a modus vivendi has evolved in which the legislature does not attempt to enforce specific directives on the university but is content to prescribe broad policies or to give broad hints of its special concerns, as through requests for reports on particular topics.

With certain exceptions, the regents are appointed by the governor.

Originally lasting sixteen years, the term was shortened to twelve years in the 1970s. Such lengthy terms ensure that no one governor will appoint the entire board. Service on the board is considered prestigious and is seldom declined. Nominally, the state senate is required to confirm the gubernatorial appointments. In practice, only one such appointment has ever been rejected. Leland Stanford, the railroad tycoon, who later endowed Stanford University, was so disliked in California that his nomination was rejected. In addition to the eighteen appointed regents, there are five ex officio regents—the governor, the lieutenant governor, the speaker of the assembly, the state superintendent of instruction, and the president of UC. There are also three regents who serve one-year terms, two UC alumni elected by the alumni, and one student chosen by the regents from student applicants.

The regents meet in the middle of each month nine times per year (omitting April, August, and December), usually for a day and a half. Most of the work of the board is conducted in its several committees—buildings and grounds, educational policy, hospitals, investments, laboratory oversight—which meet on the first day. By law, these meetings are open to the public except when specific legal, investment, or personnel decisions are discussed. The press is present, as are university personnel, parties interested in particular decisions, and, on occasion, hundreds (if room is available) of student activists agitated about one or another issue. As a consequence of this audience, the meetings at times take on aspects of amateur theater. Board members with definite political constituencies may use their positions to clearly play to factions in attendance.

At times, I, as a scientist, found it difficult to take these proceedings seriously. One of the burdens of the chancellorial position is that one is expected to consider thoughtfully on various occasions sincere but shallow experiential certainties of regents, the blatantly hypocritical certainties of local politicians, convoluted and naïve certainties of faculty, or self-absorbed certainties of callow students. To a scientist familiar with the structures of proof, these claims were often risible. But I was expected to respond to arrant nonsense with sober logic, in a dignified manner, never losing my "cool" or concern. Only later within a small discreet group could I smile, laugh, or even, resignedly, mock.

Chancellors are expected to attend regents' meetings so that they are acquainted with university policies and the reasons for their adoption. Chancellors do not sit with the board, but their absence is definitely noted. They may be called on to address the board on issues relating specifically to their campuses. A chancellor can request to speak informally to the board if he feels he has special insight concerning the issue under discussion. At most meetings, though, a chancellor plays a purely passive role. The real value of the meeting for me came in the opportunity for informal discussion with regents in the corridors during breaks, at lunch, or at the regents' dinner, to which the chancellors (and often their spouses) are invited. On these occasions, a distinct effort is made to create a sense of fellowship in service to the university, as well as a sense of pride in its history and accomplishments. The regents' meetings also afford an opportunity to talk informally with the higher officials of the central university administration, who are always in attendance as well.

Each meeting is presaged by the arrival of several pounds of proposed motions, position papers, reports, the minutes of the previous meeting, and the forthcoming agenda. Individuals or representatives of organizations may request permission to address the board on a particular topic when it is discussed. Such requests must be made in advance in a defined manner and, if approved, are for a specific duration, usually five minutes.

> Regents, regents, you can't hide.
> You believe in genocide.

Most of the business of the board is routine: approval of the annual budget; approval of building design and construction contracts; receipt of periodic reports on enrollment, student performance, affirmative action statistics, gifts, investment performance, hospital finances, and so on; discussion of new academic programs; changes in programs of faculty or staff benefits. But some issues were not routine and required special, often intense consideration. And some issues, ones in which students challenged board policy, arose repeatedly, giving rise to near-riotous conditions. Thousands of chanting students, their mob behavior belying their academic credentials, would create an atmosphere akin to siege. When students in the audience created unceasing disruption, the meeting had to be adjourned either to another room or, if none was available, until the original room could be cleared. The meeting then resumed with a restricted audience.

Divestment, until the resolution described above, was such an issue. Another has been the university management of the "weapons laboratories" at Los Alamos and Livermore. These laboratories are responsible for the design and testing of all nuclear weapons. The

University of California is responsible for their "management" under a contract with the Department of Energy. This arrangement began during World War II when the Los Alamos laboratory was established to develop the first atomic bomb under the aegis of E. O. Lawrence and headed by Robert Oppenheimer, both faculty members of UC Berkeley. Since this was to be a civilian laboratory, it was natural then for the University of California to be asked to provide its management. After the war, the arrangement simply continued, with the Livermore laboratory subsequently added. The contract has been renewed at five-year intervals.

"Management" of these laboratories, however, has been primarily housekeeping. The university provides the administrative services, the bookkeeping, the retirement program, and so on. Determination of the *programs* of the laboratories, the budget allocations, and questions of classification of research are governmental decisions. The university has a voice in the selection of the laboratory directors but not much else.

During the Vietnam War period, questions arose as to the propriety of this university role. The "Zinner committee," composed of faculty, after a review of the university oversight of the laboratories, recommended that UC should have a significantly larger role in their programmatic direction or should not continue its contract. The report had little effect. A new scientific advisory committee to the UC president was established, to be concerned specifically with scientific and technical competence of the laboratories' research, not with program or policy. The federal government was not about to relinquish nor share the selection of programs or allocation of funds.

This issue came to my attention at contract renewal time in 1978. Again, there were major student protests against this university involvement. I agreed with the students. It seemed to me to be inappropriate for a university dedicated to freedom of inquiry and the value of reason in human affairs to continue to "manage"—that is, do the housekeeping for—secret laboratories dedicated to the design of weapons of mass destruction. I did not question that the United States at that time needed these laboratories but rather felt that other agencies could serve this role as well as the university, which had only an historical connection. For the university, this role was so contrary to its professed principles as to breed a deep cynicism and distrust among its students.

The regents scheduled a session at which anyone could have five minutes to comment on the issue. I requested five minutes and, to the regents' evident surprise, spoke against renewal of the contract. Clearly, some regents had concerns about a chancellor—an administrative employee—taking such a public position on an issue that was assuredly their prerogative to decide. Others, however, opposed to contract renewal, were pleased to hear my statement. Saxon, perhaps because of his background in physics, favored continuation of the management role. However, he defended the right of a chancellor to hold a different view, particularly one based on a concern for the welfare of the university. Initially, I was the only chancellor who favored dissociation from the laboratories, but over the years several other chancellors came to favor my position.

A majority of regents, however, favored renewal, partly because of the historical connection, partly because they perceived it as bringing prestige to the university (at least within a certain sector of society), and partly because they perceived that the willingness of the university to fulfill this role engendered a favorable disposition toward UC in Washington, the major source of research funds. Indeed, some have viewed the Lawrence Berkeley Laboratory, a major bulwark of research at Berkeley, supported by DOE funds and doing no classified work, as the quid pro quo for the weapons laboratories management.

This issue recurs every five years. At the next cycle, Governor Brown indicated that he was opposed to this role of UC and would vote against it as a regent. As usual, he did not follow through. Had he exerted pressure on his regental appointments, he could have mustered the votes to defeat renewal over Saxon's advocacy. However, Brown did not do so and it was renewed by a narrow margin.

President Gardner, neither a strong advocate nor a strong opponent, seemingly has been content to let the board decide this issue. Gardner has, however, expressed concern about the amount of administrative time and attention required for this one matter, which is really rather peripheral to the primary functions of the university. In 1989, a UC faculty senate committee, the Jendresen committee, issued yet another report advocating nonrenewal of the contract. Faculty votes relative to this recommendation were subsequently taken on all UC campuses. On every campus, these resulted in majorities in favor of nonrenewal. Despite this expression of faculty opinion, President Gardner, after again making some cosmetic changes of advisory committees, recommended renewal and the board concurred. It seems likely that the regents will continue to favor renewal of the relationship unless it should be opened to competitive applications from other organizations.

Regents come to the board with varied backgrounds and interests. Many begin their terms as "single-interest" regents determined to influence the position of the board with respect to some issue of great personal concern and paying little attention to its other responsibilities. Thus, minority or women regents may initially focus on matters of affirmative action, an architect on buildings and grounds, a financier on investments, and a lawyer on legal problems. Some may have particular educational axes to grind. However, for most, though not all, regents a socialization process occurs. As they become aware of the complexity and diversity of the intrinsic concerns of the university and the forces impinging on it, they realize the interactions, the trade-offs, the compromises that must be made, continually reviewed, and revised in the light of changing circumstances. They also come to realize the value of precedent in conferring stability on this vast enterprise. In this regard, long-term regents such as Edward Carter, who was a regent for more than thirty years, serve as an invaluable institutional memory. Of course, precedent can also be used as an excuse for stagnation.

In addition to the president, two other major officers report directly to the board of regents: the treasurer and the general counsel. The treasurer in effect operates a small but professional investing team under the direct guidance and policy direction of the investment committee. The regents have been exceptionally successful in their investments and Treasurer Herbert Gordon, a quiet, conservative fellow, is highly skilled. That the general counsel reports directly to the regents and not to the president of the university is an anomaly. The president, who frequently needs legal advice, has no lawyer but must make use of the regents' lawyer. This arrangement works as long as they interact well. However, while most often the interests of the regents and the welfare of the university coincide, this is not always true. The general counsel's primary obligation is to defend the interests—the purse, the pre-rogatives, and the reputation—of the regents. He must accede to their directives even if against his own judgment.

I have known two general counsels. Both were able lawyers but were afflicted with institutional timidity. Since the university is always the object of any suit, campuses must use the office of the general counsel for any legal involvement. Consequently, the office has a large legal staff of varied competence, with some lawyers specializing in particular fields (e.g., real estate), and others assigned to handle the legal affairs of a particular campus or two. During my chancellorship, the general counsel's office, as perhaps befits the defender of the regents, was highly conservative in its approach to legal questions and had limited concern for what seemed to me to be matters of important academic principle. They much preferred settlement to trial, even in suits that seemed to me to have little merit. They did not want to trust the issue to a judge or jury. And, to be fair, they had to consider the probable cost of the litigation if the matter went to trial. However, I had to be concerned about the cumulative effect of these decisions, as it became known that essentially frivolous suits could achieve a rewarding settlement.

Nor was the general counsel willing to entertain any aggressive suit by the university to discourage baseless claims or even to seek to enforce contractual arrangements with public bodies such as the Santa Cruz County Board of Supervisors.

Here, the scientist's concern for substance—in this case, justice—conflicted with the lawyer's pragmatic understanding of legal practice and possibility. In a deeper sense, it was a conflict between two systems of truth—the scientist's "truth" based on reproducible fact and the lawyer's "truth," the evidence a jury will or will not believe—and even a question of whether it would be worth the effort to reveal that "truth." In frustration, I often wished I could hire Melvin Belli for the Santa Cruz campus.

In its rigid conservatism, the office of the general counsel turned simple legal documents into tortured tomes. For instance, in the course of negotiation with the University of Hawaii to locate the ten-meter telescope on Mauna Kea, we needed a land lease for the site. The University of Hawaii submitted a three-page lease. This was sent to the office of the general counsel and was eventually returned as a twenty-four-page document. When it was sent back to Hawaii, their lawyers joined the game and the eventual lease exceeded forty pages.

A more amusing episode concerned the reaction of the board of regents to legislation requiring them, and all university administrators, to file annual financial statements, presumably to allow detection of possible conflicts of interest. The regents were highly indignant at this invasion of privacy and breach of dignity and mandated that the general counsel sue for their exemption from the legislation on some grounds of constitutional immunity. The general counsel demurred, saying that he did not think such action had much chance of success, but the regents insisted. He filed suit and lost, then appealed—again at the regents' insistence—and lost. The regents, their honor satisfied, acquiesced and the issue vanished from sight.

The university, with five medical schools, operates five teaching hospitals. Their problems, and especially their financial difficulties and potential liabilities, occupy considerable regental attention. A major contretemps erupted at the UC Davis medical school in 1982, when the cardiology faculty accused the surgical group, doing heart transplants, of incompetence. The university was already under suit for malpractice by relatives of patients who had perished in such operations. The transplant group then sued the cardiology group for defamation. This internal melee made defense against the malpractice claims very difficult, to say the least. In addition, at this time the hospital was threatened with loss of accreditation by the state. The regents were sorely vexed, but eventually all of the legal complications were resolved by expensive settlement, including a major compensation to a transplant surgeon who agreed to move elsewhere.

A remarkable instance pitted the board of regents against the Board of Supervisors of Orange County. The UC medical school and hospital at Irvine had an agreement with the county to accept and treat local medically indigent patients. In return, the county would reimburse the hospital for its costs. If the county thought the reimbursement claimed for a case was too large, it could challenge the bill; the issue would then be resolved by an arbitrator. After a few years, the county balked at the ever-increasing total reimbursement claim, and county officials simply decided to challenge every claim UC filed. The arbitration process was completely overwhelmed. Within a year, thousands of unpaid claims had accumulated and were continuing to accumulate. Meanwhile, the university had already incurred the costs and had to pay its bills.

Letters from the board of regents to the county and meetings of a subcommittee of the board with county officials were to no avail. The board of supervisors was adamant and, in effect, thumbed its collective nose at the board of regents. This stalemate continued until the election of a new Orange County supervisor with whom President Saxon was personally acquainted. Remarkably, the two of them were able, with mutual trust and quiet diplomacy, to resolve quickly this rancorous dispute. The episode well illustrates how the prestige of the regents is not accompanied by comparable political power.

Relations between the university and the agricultural sector of the state were multifaceted. Research at the Agricultural Experiment Station at Davis and the Citrus Experiment Station at Riverside has made major contributions to California agriculture, developing new varieties of crops, techniques to improve plant yields, better pest control procedures, improved animal nutrition, and improved and novel farm machinery. The last, however, became a source of major controversy. New harvesting machinery, in particular a tomato harvester—which required the parallel development of hardier tomatoes—displaced thousands of migrant farm workers. The new machinery also hurt small farms because it was particularly beneficial to large farms (agribusinesses) that could both afford and make fuller use of it. The United Farm Workers' Union and the California Rural Assistance League and their legislative allies belabored the university over this issue. In response, the regents held extensive and clamorous hearings on the matter but did not change university policy. Dissatisfied with this outcome, the California Rural Assistance League brought suit against the university, charging it was misusing federal agricultural research funds to benefit a few large agribusinesses. The suit was fought in the courts for many years and was finally settled after the judge had given a partial ruling in favor of the university.

This issue exemplified clearly the profound sociological effects that can accompany technological innovation. In this instance, some provision should have been made, by some agency, to provide retraining for the thousands of farm workers displaced by the new machines. Similar, even more pervasive consequences may be anticipated in the future from the development of biotechnology.

UC Santa Cruz had no direct involvement in this matter, although I was indirectly caught in the commotion and furor at regents' meetings. However, Santa Cruz did have an experimental "farm"—a portion of the campus that was used by students in the late 1960s and early 1970s to grow crops as part of the "back to the Earth" movement of the times. (It was likely also used to grow certain illegal crops.)

By the late 1970s the farm, no longer fashionable, was near extinction. I was persuaded by our environmental studies program that it could be used as a model farm to study the possibility of "sustainable agriculture"—that is, agriculture that combining certain crops, farm animals, and even fish ponds could operate with a minimum of external fertilizer and also, by appropriate combinations of plants, minimize the need for herbicides and pesticides. I was favorably disposed to this proposal but insisted that it not be an ideological toy but a true research enterprise with planned experiments and carefully measured and recorded results. We were fortunate to obtain Stephen Gliessman to join the faculty to lead this agroecology project. One result of the tempest over the small-farm issue was a much more favorable attitude within the UC toward such "radical" agricultural experiments and on campuses other than Davis or Riverside.

While in general regents show each other mutual respect and tolerance, on occasion relations can become intensely acrimonious. The most extreme outburst I witnessed occurred near the end of Regent Campbell's first year as chairman of the board.

The chairman of the board chairs the meetings, calls on speakers, and, with the gavel, has significant influence on the conduct and even the outcome of discussions. The chairman is elected by the board and traditionally serves for a two-year term. Campbell, a conservative regent, then director of the Hoover Institute at Stanford and an advisor to President Reagan (who as governor had appointed him to the board), became chairman in July 1982. By this time, Governor Brown had appointed a considerable number of liberal-minded regents. In their view, Campbell chaired the board in a blatantly partisan manner, skewing the agenda, cutting off discussion, and repeatedly favoring those who shared his political persuasion.

A revolt brewed. The liberals also realized that, with the end of Brown's term approaching, their numbers might be at a peak. They declined to support Campbell for reelection for the customary second year. To soften the blow and, they hoped, to reduce the likelihood of future incessant internal conflict, they did not propose a staunch liberal but rather a well-liked middle-of-the-road regent, Yuri Wada, to replace Campbell. The older, conservative regents, however, were infuriated. The ensuing violent debate involved personal denigration, aspersions on the abilities of Wada, and language almost never heard in an open regents' meeting. In the end, the liberals had the votes and Campbell was replaced. He was so enraged that he did not appear at regents' meetings for the following year.

With Governor Brown's replacement by Deukmejian, the pattern of appointment of conservative-oriented regents resumed and the board returned to its more traditional style of conscientious but sober and genteel public service.

At the conclusion of a chancellor's service, he is given the opportunity to address some appropriate remarks to the regents. At my farewell meeting, I gave them "three wishes." I wished them good luck in their unending quest for resources for the university—a strong economy, a favorable governor, a well-disposed legislature. I also wished that, in the fullness of time, they would find a graceful path to extricate themselves from the management of the nuclear weapons laboratories. And finally, I wished them the wisdom to come to see the campuses as members of a family rather than as divisions of a corporation. I think they understood.

25

The Faculty

In his wise and delightful book, *The Academic Tribes,* Hazard Adams set forth six principles of faculty-administration policy:

1. The Diffusion of Academic Authority. . . . No one has the complete power to do any given thing.
2. The Deterioration of Academic Power. . . . Real academic power deteriorates from the moment of an administrator's first act. . . . All administrators play a zero-sum game.
3. The Diminishment of Organizational Allegiance. . . . The fundamental allegiance of the faculty member is to the smallest unit to which he belongs.
4. The Third Law of Academic Motion. . . . To every administrative action there is an equal and opposite reaction. . . . Faculties are after all composed largely of people who like problems, perhaps even more than solutions, and even to the point of actively seeking them where they have not been recognized.
5. The Protective Coloration of Eccentricity. . . . Eccentricity is not only to be tolerated in academic life; it is often a positive virtue.
6. The Necessity of Symbolism. . . . Faculties demand the proper maintenance of the symbols of their institution. . . . The president [must] . . . *seem* to exercise authority. . . . The president must therefore be a role player, a supreme actor.

In a deep sense, the faculty members *are* the university—and they know it. Their quality and dedication determine its quality and its vitality. Only they can provide the intellectual capital and the spirit of intellectual adventure for which students come and thereby create the university. The buildings, the resources, the administration are needed primarily to foster the work of the faculty. They are necessary but in no way sufficient. Their lack or misuse can inhibit the faculty, but buildings, resources, and administration can never substitute for its essential creativity. In today's university, many forces act to channel that creativity into highly specialized pursuits. No overarching intellectual synthesis unites the faculty; only procedural dogmas such as "academic freedom" claim broad allegiance.

The chancellor's relations with his or her faculty are crucial. The chancellor can provide a broad vision, but only the faculty can implement it, so he or she must work with the faculty to encourage its ideas, foster its elan, and earn its respect through fairness, integrity, and good judgment. If the chancellor and the faculty can work together, much can be accomplished. If they work against each other, wheels can only spin.

Unfortunately, a chancellor, especially one who comes to the campus from outside, can never really make close friends within the faculty, even with those members with similar professional interests. For, ultimately, he or she is their judge, the arbiter of their advancements and promotions, and the evaluator of their requests for resources. Conceivably, if a chancellor had been in an earlier time a close colleague of a faculty member, such a relationship might persist, although then he or she would need beware of the appearance of cronyism. As a result, the chancellor's position on campus is inevitably somewhat aloof from the faculty—and lonely.

For the Santa Cruz faculty, the university was still much the "ivory tower." Lacking professional schools (and excluding the small cadre that dabbled in local politics), it had limited contact with the "real world" of industry, commerce, and agriculture, of politics, crime, and wars. The faculty had largely been recruited during the late 1960s and early 1970s during the period of rapid growth of the campus. As a result, and because most had come as junior faculty, it was relatively homogeneous in age and slanted toward the relatively radical ideologies of that period. Among these was a strong dislike for authority. Some said the Santa Cruz faculty had a "talent for regicide." The failure to recruit a faculty with a more balanced age distribution and the resultant dearth of seasoned faculty with more knowledge of university traditions was another of the founders' early mistakes. Unless there could be renewed growth in the coming years, since there would be few retirements, Santa Cruz would have an aging cadre of faculty that would not be renewed by the steady influx of younger members.

Individually, each faculty member is an expert in a necessarily narrow field of knowledge and is often correspondingly ignorant of other, even closely related disciplines. Astronomers usually have scant knowledge of biology and vice versa, as do economists of anthropology, philosophers of linguistics, and Shakespearean literary scholars of Shakespearean acting. But disciplines can be clustered into broad groups with more-or-less common values and attitudes toward knowledge and perspectives on reality.

At Iowa State, I had been a relatively junior faculty member, largely immersed within my department and related fields and with very limited contact with faculty beyond science. At Caltech, a small institution, I had made contacts broadly across the institute—but the Caltech faculty has a restricted breadth of interests, encompassing primarily science and engineering, with only a small contingent of humanists and social scientists to provide a limited representation of those fields of knowledge. Now as chancellor I had the opportunity—and necessity—to interact intellectually with faculty across the broad spectrum of scholarship. This was an interesting challenge.

The scientists' values and perspectives were familiar to me. Scientists search for the single truth they believe to be present in nature. They are in essence problem solvers driven by curiosity.

A meaningful insight into an academic discipline can be obtained from the pattern of mutual evaluations within that field. Because the natural sciences have an external standard of reality, evaluations of an individual's ability to reveal that reality tend to distribute closely about a single value.

I found that social scientists, who are concerned with a more ephemeral and less tangible reality, with little opportunity to observe much less perform reproducible experiments, are often not sure there is a single accessible "truth." Frequently, they are motivated by a desire to remedy what they personally perceive as societal ills. This can unfortunately, and not infrequently, lead them to positions of advocacy and even proselytizing in the classroom, practices that are contrary to the purely educational aims of the university. (They will respond that there are no "nonpolitical" aims.) Correspondingly, evaluations in the social sciences tend to be more diverse and reflective of differing perceptions of the validity of their insights, although still centered on a single estimate.

The humanists, still grappling with the age-old and extensively analyzed issues of philosophy and literature, tend to exalt grace of expression and insight into the evolving "human condition" in these swiftly changing times. However, often they personally find a kind of refuge from these unresolvable problems in the narrow exegesis of a particular author, poet, or historical figure. During my chancellorship, valuations in the humanities were often bimodal, reflecting different schools of interpretation, with the implicit conclusion by members of one school that all members of the other were dolts.

And the artists, while admiring execution and grace of result, value above all raw creativity—the act of creation and its novelty. As a consequence, in the arts (except for music) almost "anything goes." The artists at Santa Cruz seemed quite unwilling even to formulate any rankings of other artists and evaluations were almost randomly distributed.

It is hardly surprising that such a diverse collection of intellects will find great difficulty in framing common educational goals. For the most part not broadly educated themselves, they can neither conceive nor achieve a curricular synthesis. As a result, each department can develop a curriculum for its majors, and each division can provide for the basic courses common to two or more of its departments. However, the lip service paid to the virtues of liberal education, as providing intellectual synthesis, is fulfilled only by provision of a cafeteria of "general education" courses, offered at a low level by the departments, from which each student may select any one of numerous combinations.

The colleges at Santa Cruz, each composed of a variety of faculty member cutting across departmental and divisional lines, were proposed to provide broad courses bridging disciplines and forging intellectual synthesis. While many of the original faculty who came to Santa Cruz were attracted by this abstract concept, the vision was soon attenuated by the powerful pull of disciplinary incentives, the complete lack of the additional resources needed to mount such programs, and the strong inclination of many faculty toward intellectual concentration. As well, for a faculty member with professional ambitions, development of a reputation within a discipline was a portable asset. Development of specialized, interdisciplinary courses adapted to Santa Cruz would bring only local and nontransferable acclaim.

As I have indicated, by the mid-1970s the debilitating internal conflict had eliminated most attempts at interdisciplinary courses and had left stagnation. Breaking away from the old vision released the energies of the faculty anew and actually stimulated a revival of the broad collegiate "core" courses. The collegiate structure, costly and built into the foundations of Santa Cruz, remained (and remains) as an ever-present, often tantalizing reminder of the unfilled need in higher education for intellectual synthesis, and was thus a continuing, usually low-level source of tension and occasional bickering on the campus. The faculty attracted by the initial concept were strongly committed to the importance of undergraduate education, while recognizing (in most cases) the value of graduate and professional education, which must be integral to any UC campus. And this quality has persisted as a significant criterion in the selection of new Santa Cruz faculty, much to the benefit of undergraduate teaching on the campus.

The junior faculty with which Santa Cruz began were promising but untried. Then the tenure reviews after seven years were plagued by the ambivalence of the conflicting board-college criteria. As a result, the quality of the Santa Cruz faculty when I arrived was spotty. There were some centers of real strength in astronomy, physics, social psychology, the history of ideas (called "history of consciousness" at Santa Cruz), literature, and theater arts. And there were very talented individuals in many disciplines. But it was after the "reorganization" that the criteria became much clearer and the standards for hiring and tenure rose. During my term of office, excellent groups of faculty were assembled in seismology, molecular biology, anthropology, economics, education, and linguistics. The establishment of our first program in engineering, computer engineering, proved very successful and set a standard for future programs.

We also made mistakes, even with more senior appointments. This was particularly true with deans and directors whom we brought in from outside. Academic administrators should come from the faculty, so that they will have a true understanding of academic values and faculty problems. But, unlike industry, academia has almost no systematic programs to groom and train its future administrators. Faculty in general look on university administration as an unpleasant chore that they will perform for a few years before happily escaping back to the scholarly activities they enjoy. Too long a period of administrative service makes return to studies difficult, and only a few find that they have a talent for administration or like it. As a result, when one is seeking vice-chancellors, deans, or directors, frequently there is no one on the campus qualified or willing. Then one looks elsewhere.

The idiosyncracies of Santa Cruz militate against a successful administrative transplant. Worse, as I learned by painful experience, other universities have no conscience about "unloading" ineffective administrators and will grossly inflate their abilities and minimize their defects. This is especially problematic because once someone comes on board as an administrator and proves inadequate, he or she is not likely by that time to prove a good faculty member.

Most faculty are hard-working professionals, devoted to their teaching and their scholarly activities. As a collegiate enterprise, the faculty should play a major role in the governance of all academic matters through the academic senate and its numerous committees. This requires time and effort. Some give this service reluctantly as a duty, but others see the senate as an arena for their political urges.

Faculty politics is one of the less felicitous aspects of university life. Largely eschewed by the scientists and artists who are more concerned with advancing their research or creating their works, the senate is usually the province of the social scientists and humanists. Many social scientists—sociologists, social psychologists, anthropologists, political scientists, economists—are committed to an ideological agenda such as socialism, feminism, or environmentalism. These seek especially to use the university, with its access (from a position of authority) to students in their formative years, to create attitudes that will implement their agendas in the future. They also seek to create within the sheltering cocoon of the university a model of society as "caring," "free of discrimination," and "egalitarian," the fulfillment of all the slogans that they would like to see adopted in the world at large. Some social scientists also view the university as a laboratory within which they can test their political stratagems and means to implement change. For this purpose, they may enlist the support of likeminded students to exert greater pressure on the administration.

The humanists, believing strongly in the power of the word, see the university as the place to exemplify such power. Believing that they are the true bearers of culture, but recognizing their relatively minor role in the larger society (at least in America), they tend to regard the university as their "turf," although they recognize that quantitatively they must now share it with those less cultured newcomers, the scientists. The floor of the academic senate tends thus to become the playground for a small number of activist social scientists and humanists to exercise their particular talents. The diversity of faculty interests ensures that little of consequence is passed. If proposed legislation of major importance is likely to be successful, the scientists stir from their laboratories and the artists from their studios to participate. Thus, issues of curriculum or grading policies may be debated endlessly, with little action. Intervention by the chancellor—which I attempted on behalf of more specific campuswide general education requirements—is of limited influence. The cycle of the academic year imposes further limitations. Unless a motion can be brought up, analyzed, debated, and acted on within nine months, the summer interval, like the river Styx, erases memory and one must start largely anew with new committee members the next fall. The pace of academia is best described as "glacial."

Self-satisfying and resounding speeches aside, the work of the senate is performed in its several smaller committees. The chancellor or appropriate vice-chancellor meets periodically with these committees to discuss issues of common concern, to resolve disputes, and, more generally, to keep each other informed as to larger issues that may be affecting the campus and their thoughts for future actions. The university is properly a collegiate enterprise. Its operation thereby requires extensive consultation with mutual regard, good humor, and steadfast determination if any change is to be effected.

While faculty members often make their presence and positions very evident, they can also on occasion virtually disappear. The faculty members at Santa Cruz took their mentoring role seriously. As a consequence, they were most reluctant to take any position that might alienate their students or reduce their rapport. In fact, some seemed almost intimidated by the students.

While individual students might complain about a particular faculty member, generalized student protests—whether over a tenure decision, a land-use decision, a regental policy, or perceived inadequacies of affirmative action, student aid, or concern for undergraduate education— were directed overtly at the chancellor and not the faculty. Behind the scenes, nearly everyone of these protests had some clandestine encouragement by at least a few faculty members. But the faculty members in general stood aloof, admitting no complicity, accepting no responsibility, and allowing the chancellor "to take the heat" regardless of their personal feelings on the issue, which they might convey to me privately but never publicly. Hazard Adams forgot to mention that chancellors are also expected to serve as "lightning rods."

The burden of regulation and litigation grew continually during my tenure. I doubt that fiscal scandals were more frequent than earlier, but each ratcheted upward the financial controls and auditing requirements considered necessary for "accountability." The waves from the civil rights movement of the 1960s produced ever-expanded zones of "non-discrimination"—to the classical categories of race and gender were added ethnicity, religion, physical disability, age, veteran status, and sexual orientation. Coupled with the obligation to ensure that no discrimination on these bases ever occurred were complex restrictions on the right of the institution even to inquire into an individual's inclusion in one or another of these categories.

Beyond nondiscrimination, there soon evolved an institutional obligation for "affirmative action"—positive actions to foster the enrollment and education of students from "underrepresented minorities" and to employ and advance minority and women staff and faculty members. Appropriate mechanisms had to be developed to promote these goals and to continually monitor their achievement. While the objectives of these programs were clearly desirable, to continue to uphold the overriding criterion of merit while negotiating these procedural thickets required skill and vigilance.

Affirmative action was and remains controversial. A priori, it is a violation of the principle of merit, a basic belief of the university. Further, since it is a violation of the constitutional principle of equal treatment, it is legally justified through tortured artifice such as the desirability of diversity on the campus and in the workplace. Some regard it as the displacement of "equality of opportunity" by "equality of result." However, I believe affirmative action for student admissions is justified by a closer examination of the meaning of equality of opportunity. Minority groups have historically been disadvantaged, even enslaved. Few can doubt that inheritance—cultural as well as monetary—plays an important role in American society. Compensation for past injustice, injustice still reflected in the early life of today's youth, is a valid social strategy. And affirmative action for staff and faculty—a determined search for qualified minority men and women to serve as role models for minority students—is similarly justified.

For how long? To what extent? These are tactical questions, but I believe the principle is sound. It is also, in a public university, a pragmatic necessity. A university supported by the taxes of all of the people of the state must make a serious effort to serve all of the people of the state. To the degree that the university is the door to economic and cultural opportunity and leadership, it should be open as far as possible without excessive dilution of its basic functions.

However, I do not believe that all histories are of equal value, that all cultures are equally advanced. Four years of university education is a limited span and a curriculum must of necessity be highly selective. I have never been convinced that programs in African American studies, Chicano studies, Native American studies, women's studies, gay studies, and so on should displace the study of the cultural landmarks that underlie our Western civilization. Rather, the perspectives of African Americans, Chicanos, women, homosexuals, and others should be incorporated into the broad basic courses of literature, history, and sociology. This avoids cultural fragmentation. By permitting comparison of the contributions of these subcultures with the more standard works of excellence, recognition of their distinctive features is actually enhanced. And in an increasingly multicultural society, it is important that all students acquire an understanding of the varied cultural backgrounds of the diverse ethnic groups, but not at the expense of learning the moral, legal, and philosophical bases of the American history and culture.

Happily, the faculty largely shared my views with respect to these aspects of the curriculum. They were, collectively, not as enthused about the need for faculty diversification. Affirmative action required them to go well beyond their usual networks of colleagues and established universities in their recruitment processes. Indeed, quite aside from the requirements for affirmative action, faculties often needed motivation to become more intellectually diversified. In departments such as economics and sociology, faculties tended to recruit additional faculty of like ideological persuasion. While this practice can develop strength and cohesion at the graduate level, as with the "Chicago school" of economics, it is not appropriate for undergraduates, who should be exposed to all of the principal interpretations. But apart from offering suggestions, it is difficult for a chancellor to influence a department's recruiting practices.

I soon learned that mere exhortation was not adequate. Faculty respond much more positively to rewards than to penalties. So, in addition to establishing an intensive set of monitored procedures that had to be followed in recruitment, I established a "target of opportunity" program. Each year, a certain number of new faculty positions was set aside without respect to program for targets of opportunity. These could be especially distinguished persons who, opportunely, became "available," but it was generally understood that these positions were primarily for minority and women candidates who met UC criteria without regard to programmatic considerations. These "carrots" for departments proved most effective; suddenly fine candidates were brought forward. This type of program can only be employed when the university is growing and increased numbers of faculty positions are available. However, the contacts and networks established during such recruitment remain in place and function thereafter.

On the whole, faculty behave with honor, honesty, and dignity. Regrettably, there are always exceptions, and these leave scars on the campus and on the chancellor.

Added to the roster of new rights during my term was the right to be free of sexual harassment. This freedom, essentially from abuse of power, raised particularly difficult issues as regarded its definition and the resolution of alleged instances. All of these rights have been obtained in an increasingly litigious environment that not only permits but encourages legal action by anyone who feels they have not been treated justly. With lawyers willing to undertake cases on a contingency basis and judges increasingly loath to dismiss cases as frivolous, the burden of litigation against the university has increased steadily. Today, it is virtually routine for any professor denied tenure to bring suit to challenge the decision. They have nothing to lose. The mere threat of suit by students, staff, or faculty has forced a codification and baroque elaboration of university procedures of review and discipline to a nearly stifling degree of complexity.

Some of the most distressing episodes of my chancellorship involved charges of sexual harassment brought against faculty members by students. The evidence was almost all circumstantial in these cases, with one person's word against the other's. Such charges were initially investigated by a special committee composed of faculty and staff. If in their judgment the evidence warranted a hearing, one would be held. The procedures involved were necessarily complex; confidentiality was essential as even the allegation, if public, could be very damaging. If, after the hearing, the committee believed there was substance to the allegations, the matter came to me as the final "judge" and dispenser of sanctions. It was an uncomfortable role.

In one case, a student claimed that a professor had discriminated against her academically because she had declined his proposition. The allegation was completely denied. The evidence for discrimination was, to my mind, unclear. As a chancellor should be hesitant to intervene in a professor's evaluation of a student, I found the evidence far from conclusive and, so stating, declined to take any action. Feminists on the campus were outraged. However, I believed the student could certainly have had other motivations and could have been abetted by some of the feminists, who disliked this particular professor for his rather outspoken male chauvinist views.

Another case involved a charge of actual physical assault, although not to the point of rape. I was at first incredulous because the professor had always been, to my knowledge, a likable, mild-mannered man. However, as the investigation proceeded, other allegations came forward and buried in a file was found an admonition to this same professor after a similar allegation, well prior to my arrival on the campus. In this case, then, the circumstantial evidence was cumulative and I was forced to conclude that the assault had occurred. The professor was suspended for one year without stipend.

Faculty members, of course, are human. In the two instances cited, both faculty were born in foreign cultures in which the status of women is clearly inferior, which perhaps influenced their attitudes. But the whole issue of sexual harassment is a poison to the academic scene, inhibiting what could be worthwhile mentor-student relationships. As a professor today, I will only counsel a female student in my office with the door open.

Individually, most faculty members are idealistic, generous, well-meaning (according to their own lights), and even on occasion self-sacrificing, but collectively they can blindly close ranks to repel whatever is regarded as an infringement, even a minor one, on their "rights."

During the 1980s, the issue of university-industry relations drew considerable attention. It was not a new issue, but it assumed a new importance as universities sought new sources of research funds when federal funding tightened, as the problem of international industrial competitiveness grew and with it the need for swifter technology transfer of new discoveries from university to industry, and as the potential problem area was extended to a new dimension in biology when "genetic engineering" became a reality of commercial value. In this period, the ownership of "intellectual property" such as computer programs, genetic sequences, and other "know-how" became quite valuable. Potential developed for conflicts of interest when a faculty member also became an independent entrepreneur. He might become privy to "proprietary" information that he could not share with his colleagues or his students, even if potentially useful to them. Research in his university laboratory, often funded by federal grants, might be directed toward problems of interest to his business. More broadly, whole academic research programs might be swayed by industrial liaisons.

I served for eighteen months on a UC systemwide committee to examine particularly the questions of intellectual property ownership and of potential conflicts of interest. We had discussions with faculty members who had direct entrepreneurial connections, we reviewed the policies of other universities, and we examined the relevant legal issues. Individual faculty varied markedly in their perception of the problem. Some saw it as a clear conflict of interest and took care to ensure that no research in their university laboratory was directly relevant to their commercial connection. Others were much more casual, even cavalier, about blending their academic and business activities.

These issues were clearly not simple. Faculty, students, the university, and society all had legitimate interests that merited protection, but in a manner that would further the overall goals of the research endeavor. Personally, I had some qualms about "ownership" of a basic fact of nature such as the sequence of the gene for, say, human hemoglobin. And it was evident to me that the entire development of biotechnology rested on a knowledge base developed with government funding. At the same time, I recognized the need for individual and corporate incentives, the imperatives of "venture capital," and so on.

While some universities had attempted to develop rigid guidelines (e.g., a faculty member could not also be a "line officer" in a business), we concluded that the circumstances and potential for conflict of interest were too varied to be met by a specific code. As an alternative, we proposed that all faculty regularly provide detailed reports to the appropriate dean of time spent, manner of involvement, and compensation received with regard to nonuniversity professional activity. Broad guidelines would be provided and if in the dean's judgment a potential problem existed, he or she would discuss this with the faculty member. If the issue could not thus be resolved, the dean would refer it to the chancellor, who would be empowered to take necessary action.

To our committee this seemed a reasonable, relatively innocuous way to cope with a potentially serious problem on a basis of mutual respect that assumed that all parties concerned would place the welfare of the university first in consideration. Our report was sent to the president, who referred it to the systemwide academic senate—where it quietly died. A few faculty members saw the proposal as an infringement of their "rights"; others, often remote from the areas of risk and seeing no immediate harm, respected their concerns. I fear that one day this failure to act will be a cause of great regret.

In a public university, almost all documents are matters of public record except for some aspects of personnel decisions. Faculty advancements and promotions necessarily involve judgments by colleagues. In a collegial environment, faculty members have to live and work with each other over many years, yet candor and objectivity are essential to the maintenance of academic standards. For this reason, some aspects of faculty reviews are considered "confidential" and made available only to those faculty committees and administrators directly involved in the decision process. To ensure fairness, summaries of these reviews—reworded to prevent identification of the sources of particular comments—are made available to the candidate for advancement, who is entitled to provide a rebuttal if he or she desires.

Unfortunately, on a small campus such as Santa Cruz, this confidentiality is not perfect. In particular, for tenure decisions an ad hoc committee of at least three faculty is selected by the Academic Senate Committee on Academic Personnel (CAP) together with the academic vice-chancellor to review the entire file and provide a detailed recommendation. The names of the members of this committee are considered highly confidential to avoid pressure by candidates and their friends and foes.

In one egregious instance, the names of the committee members were "leaked" by a staff person in the academic senate office to a senior member of the faculty. This person, a close friend of the candidate, proceeded to seek to influence one of the members of the ad hoc committee. This transgression was not reported and only came to light much later, in a casual conversation. Much after the fact, with the candidate having successfully achieved tenure (a likely outcome in any case), all I could do was place a strong reprimand in the transgressor's file. But I was left with the lingering concern that this instance was "only the tip of the iceberg."

The ad hoc committees play a critical role in the tenure review process and, as might be expected, do so with varying effectiveness. Some produce thoughtful and insightful analysis of each candidate's performance; others provide perfunctory reviews. In a few rare instances, the ad hoc committee report was so poor that I felt it necessary to ask the CAP to empanel a second committee. In doing so, I documented my reasons, citing the issues that the ad hoc committee had failed to address or consider.

Such an instance produced the most disillusioning episode and the most serious conflict with the academic senate of my chancellorship. The candidate, Professor S., was a female faculty member in sociology. With strong feminist views, she was quite popular with certain segments of the campus and in the local community. However, her scholarship—largely anecdotal surveys of health services available to women in the community and in prisons—seemed marginal in quantity and hardly of the quality to be expected of a UC professor. As I wrote subsequently, it seemed to me that her publications could have readily been done by "any skilled investigative reporter." In this case, the ad hoc committee report was very superficial to my mind and failed to address any of the concerns that had been raised elsewhere in the file about her work. I therefore requested a second ad hoc committee, which produced a much more detailed analysis. Subsequently, I denied her tenure. This produced an outcry and a student sit-in. Billboards appeared in town saying "Call Chancellor Sinsheimer at —— and protest the S. decision." Professor S. appealed to the Academic Senate Committee on Privilege and Tenure.

The academic senate guards its rights zealously, and the committee on privilege and tenure is regarded as the bastion of faculty prerogative. While strictly speaking its function is only advisory to the chancellor, a chancellor moves counter to its recommendations only at his or her peril. The membership of the committee is elected by the senate and varies from year to year and, while ever mindful of its mandate to protect faculty from arbitrary administrative judgments (and sometimes from each other), its interpretation of the propriety of various actions varies from year to year.

This year, the committee was chaired by a strong-minded professor of philosophy who had actually been trained in law. Popular among the faculty, he was to my mind a sophist, continually setting up straw men in his lectures whom he then demolished in defense of his quite liberal philosophy. On appeal from a faculty member, the privilege and tenure committee first decides if there are grounds for an issue. If they so decide, as they did in this case, a hearing is held.

After a quite lengthy hearing, the committee decided that the faculty member's rights had been violated—not on the issue of my judgment of her work, which they could not appropriately question, but on the procedural issue that I had convened a second ad hoc committee, which they concluded I had done in order to obtain the recommendation I wanted for "political reasons." As this procedural question had not even

arisen during the hearing, I had had no opportunity to rebut this conclusion, to point out that second ad hoc committees had been used
before in special circumstances both at Santa Cruz and at other UC
campuses. I was outraged at this gratuitous conclusion, which was to
my mind completely high-handed and a direct assault on my integrity.
I wrote a stinging letter to the chair of the academic senate asking for
an investigation of the actions of the committee on privilege and tenure.

Tactically, this was a mistake. Because of the confidentiality involved,
the facts of the case could not be openly discussed, and many faculty
simply automatically rushed to the defense of the privilege and tenure
committee, which they understandably regarded as the protector of
their fundamental rights. While the privilege and tenure decision was,
in fact, only advisory to me, as it was a direct challenge to my personal
honesty as chancellor, I decided I should refer the final decision to a
higher authority, the office of the president. This was announced at a
rancorous senate meeting. After considerable delay, the office of the
president issued its ruling, which supported my position that second ad
hoc committees were allowable under the University of California policies when such action was seemed desirable.

This concluded the matter as far as the university was concerned. By
then, the composition of the committee on privilege and tenure had
changed and in the interest of campus harmony I took no further action.
Professor S., however, went to court claiming discrimination. After
some legal maneuvering, she obtained a hearing date. The general
counsel was quite tentative about this case. He thought that the privilege and tenure decision, though it had been overruled by the president,
would taint the case and cause us trouble in court. I was more confident,
thinking we had a strong case that no university policy had been violated, as decided by the president of the university himself. The general
counsel was right and I was naïve.

The hearing was most disillusioning to me. After hearing preliminary
arguments from both sides, the judge made clear that—although it was
obvious he had not read the briefs and was quite unfamiliar with many
of the details—since there was the contrary privilege and tenure conclusion, he would rule in favor of Professor S. unless the university and
Professor S. came to a settlement. I was outraged again. I wanted to
proceed to trial and, if we lost the case, to appeal. It seemed to me
important to defend the right of the president of the university, rather
than the courts, to interpret university policy. The general counsel,
however, clearly did not feel that this was a strong case for the university

to appeal. The matter was quite out of my hands and, indeed, quite out of academic hands. The general counsel agreed to a settlement in which the tenure decision would be reopened and again reviewed by a committee of academic vice-chancellors from three other campuses. Their decision would be final. *And*, at the insistence of Professor S., the report of the second ad hoc committee would not be part of the file to be reviewed. Once again, where was Melvin Belli when I needed him?

Professor S. received tenure. I learned that: (a) the committee on privilege and tenure can have its way even when it is wrong; (b) the campus can live with one marginally competent faculty member to keep the general counsel at ease; and (c) the courts administer ideology as often as "justice."

With exceptions, a typical chancellorship in these times may cover at most a decade, whereas the tenure of a faculty member may be thirty or forty years. Thus, from the perspective of the faculty, chancellors come and go. If a faculty member does not like your policies, he or she can wait you out and hope for better. Especially in your latter years, if you are approaching a mandatory retirement age, your influence must wane on into the last "lame duck" year. And so my final year was "the calm after the storm" with the campus on a steady if noninnovative course, awaiting the naming of my successor.

26

The Students

Go Slugs!

Banana slugs or sea lions?

Santa Cruz has always had a low-key approach to athletics and I was quite satisfied to continue in a similar vein. I did not want to become involved in the corrupting programs of big-time athletics with recruiting, athletic scholarships, under-the-table bribes, and the exploitation of "student athletes" with inadequate academic preparation. I believed in strictly amateur athletics for those students who played simply because they enjoyed sport. I brought the campus into the NCAA Class III Division to play other schools of like mind.

This approach to athletics had several consequences. For one, it alienated one sector of the local community, which had hoped the university would provide them with athletic spectacles. And it is true that the provision of sports events can provide a bond between university and town, as I have observed elsewhere, and which we thus lacked at Santa Cruz. Also, because of its limited program, Santa Cruz tended to attract students who had little interest in or experience with athletics, and thus had never learned the lessons of sports, the concepts of fair play, the value of team play, the ideas that "you win some, you lose some" and "nobody bats a thousand."

The low-key attitude toward sport also gave rise to an amusing episode. When we joined the NCAA, we needed a team mascot. Some of the athletic "clubs" that preceded the NCAA teams had used the banana slug—a yellow, rather sickly looking gastropod found among the redwoods on the campus—to symbolize the campus's low regard for competitive athletics. The members of the teams, however, did not want to compete with such a mascot. A competition for mascot was held and a straw vote among the athletes selected the sea lion, a sizable mammal indigenous to the nearby California coast.

When the student body learned of this, it was indignant. They scheduled a referendum on the issue of the mascot. In defense of the sea lion, I wrote that it had "more spirit and vigor," whereas the banana slug was "spineless (ipso facto), yellow (cowardly), sluggish (slow of foot [?]), and slimy (enough said)." Of course, the banana slug won the referendum by a five-to-one margin. Acknowledging the inevitable, I accepted the outcome of the referendum: "The students are entitled to a mascot they desire and with which they can empathize. Therefore I designate the banana slug as the official mascot until such occasion as the students might wish to hold another election. I also suggest that it would be most desirable for our biological scientists to begin a program of genetic engineering research on the slug to improve the breed. The potential seems endless. Viva le (and/or la) slug." This last remark was a nod to the fact that the slug is an hermaphrodite. And so at Santa Cruz athletic events there ring out stirring cries of "Go Slugs!"

A fitting conclusion came from Chancellor Peltason at UC Irvine, who sent me a note saying, "Thank you for making our mascot [the anteater] respectable"!

The students of Santa Cruz are distinctive. During my tenure, they were largely drawn from white, upper middle class, urban California families and had had in their lives almost no direct experience of economic privation or even climatic harshness. They were a spectrum of young people, of course, and yet they were remarkably homogeneous in their political views—liberal, pro-environment, anti-authority, pro-feminist, "pro" all minorities of race, ethnicity, and sexual orientation— to the degree that there was very little political debate on the campus. Mostly, they lacked any sense of gratitude to the former generations who provided this educational opportunity; they were absent any sense of the privilege of their education as compared with the privations of most of the youth of the planet; they were antibusiness, antimilitary, at best dubious about the benefits of technology, and content to defer their entry into the "real world" for four or five more years unfettered by responsibility.

The university should be a place for study and learning, for intellectual debate and research. However, these are not the motives of many of the undergraduates. Indeed, those who pursue such paths may be labeled "apathetic." Various ideological forces (including faculty) seek to capture and exploit the idealism of youth and by inculcation to channel it toward their particular social goals. In so doing, they have politicized many campuses. By creating political tempests, they distract students from their common primary purposes and often destroy the calm reflective atmosphere essential to intellectual achievement.

I have puzzled about the sources of student attitudes and the reasons for their susceptibility to political manipulation. Affluence, of course, is one. The newness and lack of tradition in California are another. But there is, I believe, another factor as well in the late twentieth century. I grew up in a real world with real people and real necessities. During the twentieth century, we have come to live increasingly in an unreal world of music on records and life in movies and TV, all divorced from their sources, with selected fragments of sound, scene, and people in quite artificial situations. Increasingly, this unreal world has become the substance of discussion, reference, and even analysis, and young people growing up in this time expect to live in this world of boundless freedom for self-expression and scant encounter with necessity. Yet these students are also bright and questioning; many are quite talented, sure to become among tomorrow's leaders. To keep the campus on a reasonably calm course, to permit education to take place amidst the foment, actual or threatened, was my continuing challenge.

The students at Santa Cruz also have a cross to bear. Why are they not at Berkeley? Berkeley is widely regarded as the premier campus of the UC system, it receives the most student applications, and it is only seventy-five miles away. They cannot argue geography as at other UC campuses, the desire to be near home, nor are there special educational opportunities such as agriculture at Davis or oceanography at San Diego. The students at Santa Cruz, therefore, make great virtues of those features that differentiate Santa Cruz from Berkeley: the smaller scale, the collegiate structure, the beautiful setting, the low-key athletic program, the "narrative evaluation" system of grading, the greater attention to undergraduates. Reciprocally, they denigrate Berkeley as a "heartless machine" with hordes of poorly taught undergraduates, which gives its primary attention to graduate and professional education.

As a consequence, the Santa Cruz students fight any change that they perceive might make Santa Cruz more like to Berkeley. They oppose growth of the student body. Even when there were only two colleges, the students opposed construction of the third and fourth. And similarly for each addition. (In this opposition, they made common cause with the community politicians, who opposed any growth in the area.) The students oppose any policies that they feel would diminish the authority or responsibilities of the colleges versus the disciplines. They oppose any construction that they feel would disturb the natural beauty of the setting. Since cutting down trees would alter the forest and building in the meadows would damage the views, all construction, even extension of roads, is vigorously condemned. It is idle to argue that the land had been purchased to provide a campus, not a park. To minimize disruptions, we have literally resorted to such subterfuge as cutting trees in the summer when the students are away.

Expansion of research programs was resisted as withdrawing faculty attention from teaching. That Berkeley, with its faults, might nevertheless do some things well, that it might have some virtues, is an idea alien to the student culture. That beautiful surroundings and narrative evaluations are marginal features not really integral to their education is an unacceptable concept. That growth would provide educational opportunity for more students, that new facilities, more faculty research, and better cross-campus curricular integration might enhance their own educational programs are difficult arguments for them to grasp because they begin with such an animus toward "Berkeley-style" education.

I was at first taken aback by the persistence and strength of these student attitudes. Recalling the accepting nature of the students of my day—our attitude that it was a privilege to attend MIT and learn from such distinguished faculty—I was astonished at seventeen- and eighteen-year-olds who literally came out of high school convinced that they already knew precisely how a university should be designed and operated, who were determined that they would make changes, and who could not conceive of compromise.

Such students had long since seized control of the student newspaper, which—as each editorial group selected the next—had become a self-perpetuating clique. It was not really a "newspaper" but a weekly polemic devoted to fanning whatever flames could be found, and in whose pages the administration could do no right. As I came to see it, it was a steady drip of poison into the campus atmosphere. (My predecessor, Angus Taylor, had told me that he had ceased to read it, as he found it too aggravating.) Most distressing was the quality of the "journalism." I would give an interview to a student reporter only to find selective quotations taken out of context with sentences from widely disparate portions of the interview placed in direct juxtaposition to distort grossly what I had said. It was an alarming lesson in the power of the media to skew the news.

And yet, curiously, the animus of the paper toward the administration was quite impersonal. I had an annual dinner for the editorial staff that was always a cordial, even jovial affair. Individually, they were pleasant and bright kids; many of them recognized already that their attitudes would change after college. But for now they were playing out a role they felt expected of them as appropriate to their college years.

Students, of course, live within a foreshortened time frame. Any change they desire must be completed within the four or five years they are on campus or it is meaningless to them. This accounts for the often feverish character of their protests in the face of the generally glacial pace of change in academia. Minority students who want more minority faculty members now do not want to hear about the years it takes to augment the pipeline of minority graduate students and Ph.D.'s and faculty retirements before any large numbers of minority faculty will be available and in place. Ethnically, Santa Cruz and Santa Barbara are the least diverse campuses, although both are growing in proportions of nonwhite students. Located in nonurban areas, both have no large pools of nearby minority applicants. Indeed, there is simply no sizable black community in Santa Cruz and only a small Asian enclave. Their absence is a serious lack for Santa Cruz students in these groups.

Among the UC campuses, Santa Cruz students customarily rank second to Berkeley in average SAT scores of verbal ability and fourth or fifth in mathematical ability. The lack of an engineering school doubtless affects the latter score. The campus today attracts academically able students who are articulate and who hope to be able to take advantage of its less structured curriculum and greater concern for undergraduates. And many do. A higher proportion of students go on the year-long Education Abroad program from Santa Cruz than from any other UC campus. At the same time, however, the less structured programs allow less motivated students to "slide by."

The narrative evaluation method of grading is clearly favored by the students. There can be no doubt that it reduces competition between students and the incentives to cheat or even to sabotage other students' work. With well-motivated students, the absence of grades can foster the concept of a learning partnership between student and instructor. Regrettably, not all students are well motivated. Indeed, for some, the grade *is* the reward.

Outside the campus, the "lack of grades" is widely perceived as evidence of educational laxity. And many graduate and professional schools, overwhelmed with applicants, simply will not take the time to read a string of paragraphs. For these reasons, I urged the faculty to allow optional grades in all upper division courses for those students who wanted to have them on their records in addition to the narrative evaluations. Much to the students' initial displeasure, the faculty concurred. After a few years, this is now the accepted pattern. Only about 25 percent of the students eligible to receive grades request them.

When I arrived at Santa Cruz in 1977, the hippie tide was belatedly ebbing. Long hair, unkempt beards, and scruffy attire were still common, but year by year this appearance dwindled. By the mid-1980s, hair was of moderate length and (mostly) combed, clothes were clean, and beards had vanished. Some girls returned to pretty dresses and used makeup. Student interests changed in parallel. Biology, curiously, has long been the most popular major at Santa Cruz. But whereas psychology and environmental studies had been the next most popular, they were replaced at least for a time by computer and information science and by economics with a business track.

As a symbol of the campus, the chancellor is in the popular mind accountable for the appearance and actions of his students. The severe restrictions on his actual authority are not generally recognized. The university surrendered the in loco parentis role in the 1960s, when society in general ceded more responsibility to youth as in the eighteen-year-old vote and the "sexual revolution." Right to privacy legislation has even placed student dormitory rooms beyond the reach of institutional access except in case of emergency. The use of drugs, or alcohol by minors, is illegal but cannot be policed. Because of court interventions, suspension or expulsion of a student is now difficult and must follow elaborate rules of due process and quasi-legal procedures.

The legal situation of the university and the student newspaper is particularly anomalous. The university is legally the publisher of the paper and is responsible for any potentially libelous statement therein. Yet the courts have ruled that the university has no right, under the First Amendment, to censor the paper. Thus, as chancellor I was responsible for a publication over which I had no effective control. In fact, the university was twice threatened with libel suits by off-campus persons over articles in the student paper. Fortunately, in both cases the suit was ultimately dropped after tempers had cooled.

Every week, I held open office hours for students, normally on Tuesday afternoons. Any student could come in to see me without an appointment. A few students were curious about the chancellor: What was he, the reputed ogre, really like? But I soon found that most students either had a complaint or wanted special financial support for a project or journey. In coping with complaints, I quickly learned that I must work through the system. In principle, as chancellor I had the authority to reach into the system and fix what might appear to be an evident malfunction or injustice. But if I directly bypassed the chain of responsibility, I would create confusion and resentment in those whose job I momentarily usurped. The malfunction or injustice could be corrected only by passage of an instruction through the chain of authority to the proper locale of action.

In time, response to these common reasons for visits to the chancellor was institutionalized. A position of ombudsman was created to mediate student or staff complaints and to straighten out inconsistencies or contradictions in policies. And a committee was established to periodically review student requests for special funds for projects—plays, dance performances, visits to research centers or archeological digs, and so on.

To encourage and recognize exceptional student performance, I established the Chancellor's Prizes—annual monetary awards in a variety of categories, such as best student research project, outstanding service to the community, best student dramatic performance, and so on. These awards were announced at graduation.

It may seem that most of my interactions with students were confrontational. Unfortunately, this was true, but not by my choice. Each year a fresh crop of student activists felt the urge to launch into its new profession. Apart from tenure cases, it was difficult to forecast which would become the salient "issue of the year." I sought, when I could, to look on the student protests as another educational opportunity. Although the discussion might engage much of the student body, the numbers directly involved seldom exceeded a hundred. Most students were typically going about their normal studies. Sometimes students sought to educate each other in blunt fashion. One amusing (?) incident occurred when left-wing students sitting-in at the chancellor's office to protest a tenure decision called the campus police late one night because they were being harassed by a small group of right-wing students. Calling on the authorities to protect their own illegal activity reflected an almost touching innocence.

But once an issue was raised, the fragmentation of the student body among the several colleges made it difficult to ascertain, much less influence, student opinion. (For reasons already given, the student newspaper was not an accurate barometer.)

Who spoke for the students? Each college had a student council. To attempt to meet with eight student councils on every issue was a considerable task. And not infrequently, the several student councils would disagree among themselves. During the period of protests concerning divestment from South Africa, a group of students objected vigorously to the presence on campus of an automated bank teller from Wells Fargo Bank, which did business with South Africa. We had merely made space available on campus for this teller as a convenience to students and staff. No one was obliged to use the teller, but some nine hundred students were regular customers. Should their prerogatives be abridged to mollify the protestors? Some of the collegial student councils took positions on this matter and were not of one mind. I declined to order the teller removed but arranged with a local credit union for a second teller to be installed so that students could have a choice.

A similar situation arose with regard to the "cyanide pills" controversy. In 1984, students at Brown University voted that cyanide pills be stored on campus to permit mass suicide should a nuclear war begin. In copycat fashion, a similar proposal was soon made at Santa Cruz. (Modern communications make possible systemwide, nationwide, even international coordination among student groups.) I made clear that in no event would I permit such capsules to be stored on campus, but the debate raged among the students until the proposal was finally narrowly defeated in a referendum.

For this and other reasons, I favored the establishment of a campuswide student union, but this idea had to originate with the student body. Because of opposition by some of the college councils, earlier attempts to establish such a union had failed, as did an effort early in my tenure. However, a second attempt toward the end of my term succeeded and a campuswide student government, albeit with strictly limited powers, now exists. Among other functions, it provides a home for campuswide organizations such as minority student alliances and, I hope, will provide a forum in which student-administration issues can be more rationally and productively discussed.

Students graduate and become alumni. As a young campus, the alumni of Santa Cruz were all young with recent memories. When I became chancellor, the alumni were dominated by the students of the late 1960s who cherished their memories of the early heady and euphoric years of the campus, now recollected in tranquillity. Unacquainted with more recent troubles, they were reluctant to see the need for any changes in its structure. I worked with them and sought to aid and strengthen the fledgling alumni organization and to involve alumni in campus programs such as student recruitment. The loyalty of alumni is important for the prestige of a campus as well as the potential for its future financial support. It is my hope that future chancellors will benefit.

27

The Community

Many of the leading politicians of Santa Cruz after 1970 were refugees from Silicon Valley or the suburbs of Los Angeles. They had seen semi-rural areas become urbanized and fled' to Santa Cruz. And they were determined to use all means to prevent the repetition of what they had seen, to keep Santa Cruz small and laid back. They felt no obligation to the prior promises made to UC to bring the campus to Santa Cruz, nor did they feel any societal obligation to help to accommodate the continuing influx of population into California.

They found ready allies in the voting student body, strongly influenced by the environmentalist movement and entranced by the natural beauty of the area, which contrasted with the urban settings in which most of them grew up. They were also aided by the relative apathy of a growing "bedroom community" who commuted daily "over the hill" to jobs in Silicon Valley and had less concern for local affairs. The no-growth policies had the reinforcing effect of raising property values. However, such policies had a negative economic effect, discouraging the entry of business into the area and seriously limiting employment possibilities. Young people growing up in the area eventually had to leave to find career opportunities.

During my tenure, relations with the political figures of the local community were an unending confrontation. Contrary to my expectation, they showed no appreciation for the social and cultural benefits provided by the presence of a major university. Politicians must, of course, reflect the wishes of the constituents who elect them. However, one might hope for a larger vision from those who have in their hands the future welfare of the community, as well as some sense of responsibility for the commitments of their predecessors. Not so in Santa Cruz. Even the potential of the university, a "clean industry," for direct employment and secondary employment through the provision of services was well beyond the range of their limited vision.

In 1982, after four consecutive years of belt-tightening and fiscal stringency under Governor Jerry Brown and with no relief in prospect, I sought urgently for a way to improve the campus's finances. Unlike some of the UC campuses, we had no real potential to raise much money from our young alumni or from local business. We did have one major resource—two thousand acres of land in a very desirable part of California. Manifestly, it could not be sold, but it did occur to me that a portion of it could be used to house a research park as had been established at Stanford and several other universities. A research park would have several advantages. It would significantly augment the numbers of scientists and researchers, thus providing potential interaction in our rather isolated community. It would provide part-time and summer employment for students. It would provide employment for the significant number of unemployed in the Santa Cruz area, as well as a tax base for the local government. And it was a "clean industry."

I commissioned a consulting firm to do a survey of potential interest for involvement in such a research park among firms in northern California and, in particular, Silicon Valley. Their report was positive and it appeared that a research park would be attractive and feasible and would in time provide substantial revenue to the campus. I then publicly broached the proposal. To enlist support and dispel misconceptions, I met individually with all eight college faculties and all eight student councils. The responses ranged from mere acceptance to enthusiasm, but there was no organized opposition.

Then the proposal reached the local politicians, who were opposed to growth in any form. While much of the employment to be generated could make use of unemployed persons already living in Santa Cruz, we estimated that the research park, when fully developed, might bring in as many as two thousand scientists, engineers, and technicians. As one of the political leaders opposed to the proposal said to me in a rare moment of candor, he did not think they would vote for him! I was still näive enough to be astonished at such naked venality. Of course, the first and enduring principle for a politician, as distinct from a statesman, is to be reelected. These are not citizen-rulers, nor even civil servants. They are career office-holders.

Ordinarily, construction on a UC campus, as an entity of the state, is exempt from local zoning or growth control ordinances, although not from environmental control requirements. Nevertheless, county officials threatened to bring suit to force the research park to comply with local requirements (and thereby kill it) on the basis that it would not be part of the educational mission of the UC. The university general counsel thought, however, that we could make a good legal argument against such a challenge. Further, though, the construction of a research park would definitely require construction of the direct-access road to the campus long ago promised by the county when the UC located in Santa Cruz. And unless the county would build this, we were stymied.

I wanted the university to apply pressure, political and otherwise, on the county to fulfill its long-standing obligation. At this time, however, the presidency of the university changed hands, and when I discussed the whole research park project with the new president, David Gardner, he was not enthusiastic. There had been a research park adjacent to the campus at his former post at the University of Utah. And while it was clearly a successful park, he did not feel that its presence had been especially beneficial to the campus and that, rather, it had been a source of problems relating to faculty-industry connections and so on.

In short, a research park at Santa Cruz was not worth the political capital it might cost UC.

Without the president's strong backing, I had no means to exert pressure on the local officials, and so the research park project was burked. I still believe it would have been beneficial for both the campus and the community.

In 1984, application for enrollment at Santa Cruz turned upward and again, sharply, in 1985. The bad image of the campus was finally behind us. And newer demographic projections for California indicated that UC as a whole could expect marked growth into the twenty-first century. For the first time in fifteen years, we could envision the campus achieving its planned 27,500 students.

Still an optimist, I thought I should personally bring this good news to the attention of the local political leaders. When they had vehemently opposed the research park because it was not "educational," they had stated they would have no objection to growth of the student body. In fact, I thought that since the students in general supported their policies, they would welcome more student voters. I proposed a larger vision. Santa Cruz was distinctive among the UC campuses in the small size of its surrounding community. This provided a unique opportunity to create in Santa Cruz a true "university town," a community focused on its university as its employment base and cultural center, much like Cambridge in England and Princeton in the U.S.

I thought this vision would be appealing, but I was wrong again. The politicians' no-growth reflexes took command over whatever cultural ambitions they may have had for their community. They could not prevent our acceptance of more students, but to accommodate more students the campus required more buildings. These, in turn, required the preparation of a new campus Long Range Development Plan, which required an accompanying Environmental Impact Report. The latter had to detail all impacts of campus growth on the local environment, including of course the community, and to propose "mitigation measures" where possible. The city or county governments could challenge the sufficiency of these mitigation measures in court if they wished, and they threatened to do so. We might win the court case, but all construction could be delayed for years.

My retirement came just after completion of the Long Range Development Plan and before completion of the Environmental Impact Report. My successor had then to negotiate this difficult problem with the local officials. He finally obtained their reluctant acquiescence, but at the price of limiting future campus growth to fifteen thousand students. The campus could make a strong educational argument that 15,000 students was the minimum number that would, under UC funding rules, provide sufficient faculty to permit the campus to staff adequately a range of programs to provide enough breadth to be considered a major university. The local politicians felt they could not successfully challenge this rationale.

There were, of course, components of the community that appreciated the cultural benefits afforded by the campus and that supported it politically and financially. From these we drew members of the UCSC Foundation, the UCSC Affiliates, and the various friends groups for particular activities such as the arts, arboretum, library, marine laboratory, and so on. And among these we established strong personal friendships. Karen (I had remarried) recognized that while our amateur sports could not provide a strong bond with the community, the arts could provide an alternative network of common interest. With the aid of the theater arts program, she established a summer Shakespeare festival involving actors from the Royal Shakespeare Company as well as more local talent. Staged on campus but governed by a joint campus-community board, the festival grew rapidly and attracted large audiences from the community and, later, the greater Bay Area. The festival has become a major Santa Cruz attraction and its existence has significantly enhanced our theater arts program.

It seemed to me to make eminent sense to combine the resources of the campus and the community to enable us to attract a more eminent music conductor for both the Santa Cruz Symphony and the campus orchestra and a joint appointment was thus arranged.

It is so obvious that the campus and the community should make common cause that one can only believe that, in time, common sense will prevail over ideology. In the meantime, relations remain dismayingly difficult.

28

"What Does the Chancellor *Do?*"

"What is the chancellor?"
"Well, the chancellor is the chief administrative officer of the campus."
"I know that, but what does the chancellor *do?*"

The chancellor has many roles—symbol of the campus, campus representative in many settings, campus host, mediator and arbiter of disputes, target for protest, douser of "fires," and *always* mentor.

Importantly, given his or her limited authority and resources, the chancellor should constantly look for, probe for, create, and exploit "windows of opportunity" to improve the varied aspects of the university. He or she must provide the initiative for those times when faculty willingness and foresight, staff needs, student tensions, resource accessibility, and personnel availability combine to make possible a major strengthening or broadening of the academic enterprise—a new department or emphasis, a new and important research program, a plan for staff training and advancement, a student recreation center or a campuswide student government. Such times seldom occur spontaneously but must be catalyzed and nurtured amid the daily routine and (often) tumult.

It took me a little while to realize that to be the chancellor at Santa Cruz was to occupy a distinctive, if somewhat lonely, niche in the UC system. UC routinely pays lip service to the desirability of improved undergraduate education, but its heart is in graduate and professional education and research, and its pattern of resource allocation reflects this priority. As a campus that sought seriously to emphasize a high-quality undergraduate education (together with graduate education and research), Santa Cruz was hobbled by the UC ideology.

UC receives from the state a budget primarily related to the numbers of undergraduate, graduate, and professional students on a per student basis. The state customarily agrees to fund all qualified undergraduate students and a negotiated number of graduate and professional students. But UC does not *distribute* its funds to the campuses on a per student basis. Rather, it uses a "weighted student" formula in which lower division students (freshmen and sophomores) count 1.0x, upper division students (juniors and seniors) count 1.5x, beginning graduate students count 2.5x, and advanced graduate students count 3.5x. (Professional students have other weightings.) Manifestly, such a formula for resource allocation favors those campuses with large graduate student enrollments and drives all campuses toward similar emphasis on graduate programs in order to be able to compete more successfully for the limited resources.

I argued year after year against this allocation pattern, with complete lack of success. The bland argument was made that "it costs more" to educate graduate students than undergraduates. I argued in vain that this distinction depends entirely on the manner of instruction—that one can usefully spend just as much on the education of each undergraduate as UC spends on each graduate student, and that, indeed, many first-rate private institutions do so. The situation was made even more invidious by the circumstance that Santa Cruz was not allowed to improve its graduate/undergraduate ratio so as to compete more successfully with the large UC campuses. In these years, the state would not agree to increase the funded number of graduate students and the larger campuses were not about to share their allotments, acquired in earlier times, with the newer campuses.

This situation resulted in the irony that as campuses such as Santa Cruz and Santa Barbara increased their undergraduate enrollments, the resources and faculty positions that UC acquired from the increased state budget went, under the formula, in considerable part to the large campuses that had not grown at all. When UC changes its allocation formula, I will believe its rhetoric about the importance of undergraduate education.

Over the years, I sought to advance the interests and distinctive ideals of Santa Cruz within the UC system while at the same time seeking to reconcile Santa Cruz to the UC standards of quality. Within the collegial ambience of the university, significant curricular change is almost impossible except by growth. During my first five years, with ever-tightened budgets, change could only come by taking resources and faculty positions from one program to transfer to another at a time when all were constrained.

Because it had been assumed the campus would grow without pause for decades, by 1977 twenty-three boards of study had been established, all stretched very thinly to cover their disciplines. The board of religious studies was in grievous straits because of inadequate initial staffing and subsequent resignations. An external review of the program recommended either a doubling of the faculty or its termination. As the former was infeasible and the program was not a central or core discipline, I decided to "disestablish" it.

Once established, a program within UC can only be disestablished by an elaborate procedure, over a two-year period, involving numerous committees and hearings to provide opportunities for campus and other comment. Ostensibly, this procedure allows for broad input to enable the administration to decide whether it really should terminate the program. During this lengthy period, the administration is of course subject to constant pressure from those students, faculty, alumni, and colleagues at other campuses who resist the action. The prolonged process thus becomes a facade, for no administration would subject itself to such pressure and condemnation unless it had already made its decision.

During my second five years, budgets were easier, the campus was growing, change was more feasible. Working with the faculty, I sought to broaden the academic base of the campus and, with an eye to the future, select its directions of emphasis. Faculties, absorbed in their own intellectual interests, tend to reproduce themselves when thinking of new faculty. To dislodge such tendencies, I initiated a program of external reviews at five-year intervals of each board of studies. Such reviews require the board to reflect on and define its current program, and they bring an objective and informed external perspective to bear on the assumptions underlying the current programs.

With their advice, with augmented resources, and with leadership from farsighted faculty, we were able, as has been mentioned, to develop several outstanding programs. A special committee composed of engineers from both university and industry helped us to recruit Patrick Mantey, a leading researcher from IBM, to establish our program in computer engineering, our first step toward a revived school of engineering. With the committee's aid, Mantey in turn brought in an excellent faculty and established valuable ties with Silicon Valley.

Another achievement was the establishment of Phi Beta Kappa on campus. Shortly before I arrived at Santa Cruz, the campus had been turned down in its request to establish a chapter of Phi Beta Kappa because of administrative turmoil and uncertainty as to its future academic directions. This was a further blow to campus morale and imposed a "penalty" on our best students, who could not be awarded this distinction. Therefore, following our efforts to improve academic quality and having reversed the decreasing enrollment trend, we reapplied in 1984. This time, after the usual period for investigation, the establishment of a chapter at Santa Cruz was approved.

The social obligations of a chancellor are formidable. Karen and I hosted some 140 to 150 events at University House each year, mostly between September and June. Events included lunches, dinners, receptions, award ceremonies, support group meetings, and so on. At another university, many of these events might have been held at the Faculty Club, but as a young campus, Santa Cruz lacked such amenities. So the University House (the chancellor's residence) became the only appropriate locale for events of campuswide interest. Because of its ambience and facilities, many community organizations would have liked to use University House for their activities. The potential list was overwhelming and to keep it manageable we sharply restricted such use to organizations with direct university ties.

One of the fringe benefits of being a chancellor is that the university attracts or invites distinguished speakers and over the years it was our privilege to entertain them. The Dalai Lama of Tibet spoke on campus to a throng of over five thousand. We spent a fascinating afternoon with him. As the titular head of a theocracy, he is a remarkable combination of religious leader and shrewd politician. He also had a deep and abundant sense of humor.

He told us of his selection to be the fourteenth Dalai Lama. Each Dalai Lama is believed to be the reincarnation of the former. Two or three years after the death of a Dalai Lama, a "search party" is sent about the nation to find, among the children of appropriate age, the reincarnation. He is identified by his ability to recognize, specifically, favored possessions of the former Dalai Lama, a comb, a book, shoes, and so on. This Dalai Lama was located in a remote corner of Tibet. He was then, at age of about 2½, brought in an ox-cart on a three-

month journey to Lhasa. There he was educated in the Tibetan Bud-
dhist tradition and in the ways of governance. After passing tests, he
was ordained as a priest and inaugurated as the Dalai Lama. Of course,
early in his reign, the Chinese government overran Tibet and forced
him into exile. He still desires and hopes to return to Lhasa.

A very different visit was that of two Santa Cruz alumni who have
become astronauts, Kathleen Sullivan and Steve Hawley. Both were
remarkably poised, outgoing, and articulate about their experiences,
creating a great sense of pride both in NASA and in Santa Cruz as the
"birthplace" of their careers.

Carl Sagan was another who attracted an immense crowd. Far more
affable and unpretentious than he appears on TV, he made a strong
appeal concerning the need to "preserve the planet."

Gore Vidal was in fine form, witty and acerbic. At the time of his
visit in 1982, he was contemplating a run for the U.S. Senate, which
would have put him in opposition to Governor Jerry Brown. He told
us that Brown had offered him a position on the UC board of regents
if he would not run for the Senate. He declined the offer but in the end
did not run anyway. His presence would have surely enlivened regents'
meetings.

Harold Wilson, the former prime minister of Great Britain, told
many amusing anecdotes. He pleased audiences with a remarkable im-
itation of Winston Churchill. And Tom Wolfe, dapper in his trademark
white suit, engaged a large audience with a witty, frequently barbed
lecture on the decline of American letters.

"Name recognition" was the key to student audience attraction for
these visiting luminaries. Extremely able and distinguished speakers
such as Simon Ramo and Elizabeth Holtzman drew embarrassingly
small audiences.

The chancellor confers the degrees on graduating students at commencement. At Santa Cruz, each of the eight colleges has its own commencement and a ninth is held for graduate degrees. Basically, these are
joyous affairs to mark the successful completion of a demanding formal
education. But at Santa Cruz they were often corroded by the diatribes
of student speakers reaching for an acme of rebellious rhetoric in their
last fling. As several of the graduations were simultaneous, I could not
attend all nine. Usually, I officiated at four college commencements and
the graduate student ceremony; responsibility for the other colleges was
then delegated most often to the academic vice-chancellor.

The smaller-scale collegiate ceremonies permitted more individual recognition and informality, albeit sometimes at a cost to the dignity of the proceedings. In the late 1970s, attitudes of the 1960s persisted at Santa Cruz. Few students wore cap and gown, and bare feet and bizarre costumes were not infrequent. Student speakers invariably trashed the academic administration and the venal larger society into which they were about to enter, often to the dismay of the assembled parents. With time attitudes became more conventional and by the mid-1980s most students wore cap and gown and the student rhetoric was more temperate even if of similar thrust.

I repeatedly proposed a collective graduation ceremony for all of the colleges and with a major speaker as a "grand occasion." This could have been followed by separate programs at each college that would provide opportunity for individual recognition. The colleges, jealous of their autonomy, consistently rejected this proposal.

The graduate ceremony was much more restrained and dignified. A senior faculty member would provide a usually brief address. The small scale permitted reading of the thesis title as each candidate was hooded, individually, by his major professor. Each degree conferral was a personal and often touching moment.

Some of the rewards of being a chancellor come long afterward. The growth in student enrollment beginning in 1984 finally brought construction funds to the campus, but given the protracted authorization, planning, procurement, and construction schedules for public buildings, none were completed before I left. The first major new building to be completed was a large new laboratory for biology and chemistry. The regents were kind enough to waive their rules and to name the laboratory building for me. At the dedication, I indicated that I was indeed honored to have this enduring monument named for me, especially while I was still alive—unlike James Lick, I did not have to be buried under the structure. James Lick donated the funds for the construction of the Lick Observatory. Unfortunately, he died before it was completed and is buried in a crypt beneath the base of the original refracting telescope at the observatory. "I am glad that I did not have to be interred under an ultracentrifuge. Talk about turning in your grave!"

So much of my effort was devoted to raising the quality of the scholarship at Santa Cruz. I was therefore simply delighted when in the spring of 1991 the Institute for Scientific Information published an analysis of the "citation impact" of scientific papers, published in the previous three years, from academic institutions. (The "scientific impact" measures the extent to which others refer to these papers.) In the physical sciences, the papers from UC Santa Cruz ranked first among all universities in the nation; in the biological sciences, twelfth. While the specific significance of these rankings can be argued, they surely place the quality of scientific research at Santa Cruz among the best in the nation.

29

The Telescope and
the Genome

Mrs. Marion Hoffman died on Friday, 16 December 1983, a sad event that was to have remarkable consequences.

Not long after I became chancellor at UC Santa Cruz, I was made aware that I also thereby became responsible for the fortunes of Lick Observatory. Lick Observatory was established in the 1880s as the observatory of the University of California as a memorial to James Lick, an eccentric Bay Area millionaire who donated the then considerable sum of one million dollars for this purpose. (A prior notion of Lick had been to build a great pyramid rivaling that of Cheops in what is now downtown San Francisco.) When dedicated in 1888, Lick Observatory was the first major telescope on the West Coast and possessed the largest refracting telescope anywhere. Not long after, it acquired the first Crosley reflecting telescope, which established that reflectors could provide images of quality comparable to that of refractors. Throughout the years, Lick had been a premier observatory although overshadowed by larger instruments in southern California at Mt. Wilson and later Mt. Palomar. After World War II, Lick had obtained funds from the state and built a 120-inch telescope, then one of the largest telescopes in the world.

Originally, Lick had been an autonomous unit of UC. The astronomers had lived in considerable isolation on the top of Mt. Hamilton. With the establishment of the Santa Cruz campus in 1965 and the greatly increased ease of transportation, it had been decided that the observatory, while still a UC systemwide facility, should be assigned to

Santa Cruz, and that the astronomers would move off the mountain and teach and have their facilities at Santa Cruz.

But now Lick had a new problem. San Jose, once a small, semirural area, had become the focus of Silicon Valley. With the rapid increase of population came light pollution, harming the visibility at Mt. Hamilton forty miles away. To escape the glow of light in San Jose, the university was considering the establishment of a new facility about a hundred miles to the south on Junipero Serra Peak, in the Santa Lucia Mountains west of King City. Light pollution was nonexistent in this area and seemed likely to remain so for many decades. Whether the telescopes on Mt. Hamilton could be moved or whether new ones should be built was as yet an unanswered question.

Junipero Serra was on land belonging to the Bureau of Land Management. Surveys had already been made and consideration given as to how a road might be routed into the area and up the mountain. However, a major obstacle had arisen. As the highest peak in the Santa Lucias, Junipero Serra was a sacred mountain to the indigenous Native American tribes. The tribe leaders were vigorously opposed to granting of a permit to build a road and observatory. Attempts to persuade the tribal leaders of the value of the observatory, of the abstract and aesthetic quality of its astronomical uses were futile. Attempts to compromise the intrusion and proposals to minimize construction on the peak and to avoid some road construction by the use of aerial tramways and so forth likewise were bluntly rejected.

Stymied at least for the moment but stimulated by the thought of other telescopes and other sites, another idea began to take form in the minds of the Lick astronomers. If a new telescope were to be built, should they not seek to surpass any existing instrument? The largest effective telescope was the Hale telescope at Mt. Palomar, a two-hundred-inch reflecting mirror. It and its supports and machinery had been designed in the 1920s, when it represented the forefront of the technology of that day. Indeed, all subsequent telescopes, albeit of somewhat lesser size, had been modeled on the Hale telescope design.

Now, a half-century later, could not a more advanced design be conceived? A larger mirror would provide more light-gathering power. This would permit studies of more distant galaxies, out so far that their light began its journey not long after the Big Bang that originated the universe. A ten-meter (four-hundred-inch) telescope, twice the size of Mt. Palomar, would collect four times as much light. But given the laws of geometry, and the need for the mirror to bear its own weight, a ten-meter telescope based on the Palomar design would weigh eight times as much. The machinery to move the mirror with precision would be correspondingly more massive. It was estimated that a scale-up of the Palomar design to ten meters would cost about five hundred million dollars, if indeed it could be built. A technological breakthrough was needed that would permit the use of a much lighter mirror and correspondingly simpler machinery.

Two approaches were developed. The one at Lick was a single "meniscus" mirror, ten meters in diameter and very thin, whose shape would have to be precisely maintained by its supporting structures. Another approach, by Jerry Nelson at Lawrence-Berkeley Laboratory, was a mirror composed of thirty-six hexagonal sectors, each nearly six feet across, to be maintained in precise register by a series of sensors that monitored the relative positions of the edges of each sector and in turn controlled the position and inclination of each mirror through an integrating computer program. The mirrors themselves were to be formed by a new technique of stressed mirror polishing in which a circular glass blank was first appropriately deformed by peripheral weights and then polished to a spherical figure (a simple task), after which the constraints were removed to allow the blank to relax back to the proper parabolic shape.

By 1980, a decision had to be made if the project were to proceed. A committee of senior astronomers from outside UC was convened to make a recommendation. By a vote of four to three, they opted for the sectional concept. Personally, I was very pleased with this result. The meniscus concept might work, but ten meters, if attainable, was clearly an upper limit for this approach. If a ten-meter sectored telescope could be built and worked well, extension of this concept to even greater size was straightforward.

We now needed a substantial influx of funds to develop the technology to the point at which a full-scale telescope could be built. We also needed to launch a search for the necessary funding, estimated at that time at fifty million dollars, for construction. We took the matter to President David Saxon. A physicist, he immediately grasped the virtue of the proposal and the ingenuity of the concept and supported it in principle. He convened a meeting of the chancellors of the four campuses—Berkeley, Los Angeles, San Diego, and Santa Cruz—with interests in astronomy to discuss the proposition. Opinion among the chancellors was certainly not unanimous. I was the most vigorous advocate; Mike Heyman, then new at Berkeley, was supportive; Chuck

Young at UCLA was neutral or perhaps moderately disinclined; Atkinson of San Diego was clearly opposed, given the other needs for funds as he saw them.

After strong argument and with unconcealed but not leading support from Saxon, this group agreed to proceed with the project. Saxon would find funds, about one million dollars a year, to carry the design process forward. A seven-member executive committee comprising chancellors or their representatives from the four campuses, the Lick director, the LBL director, a representative of the user-astronomers, and the vice-president for academic affairs would oversee the operation and a search would begin for the needed funding. At the end of the initial meeting, Saxon recalled how in the midst of the Depression President Sproul had managed to devote one million dollars of university funds to E. O. Lawrence to build his early cyclotron, to the university's and the nation's everlasting benefit. Saxon indicated that he thought this to be an analogous prospect.

The search for funds took the form of seeking a single large donor who would be intrigued by the project and wanted a personal memorial. Eugene Trefethen, a loyal Berkeley alumnus and former vice-president of Kaiser Industries, a man well acquainted among the ultra-wealthy, agreed to seek to find such a donor either in the United States or abroad, specifically in Japan or Hong Kong. He thought there were a reasonable number of prospects and felt confident.

The technical development, largely under the supervision of Jerry Nelson at LBL, proceeded more slowly than anticipated but systematically and without encountering major difficulties. Costs seemed to escalate, but President Saxon loyally managed to find the funds needed.

The search for the single donor, however, had proven fruitless. Tax laws in Japan diminished the incentive for multimillionaires in that country. Meanwhile, more refined cost estimates had risen to about seventy million dollars.

Discussion of a possible site had fairly quickly settled on Mauna Kea, an extinct volcano on the big island of Hawaii. Three telescopes had already been located on this peak. Light pollution was negligible. At fourteen thousand feet, on an island surrounded by water at uniform temperature, studies of the lower atmospheric turbulence had indicated that the "seeing"—the stability of the optical image of a distant object—was consistently superior. Further, some of the more interesting research in astronomy now involves the study of light received in the infrared region of the spectrum. Certain portions of the infrared are, however, absorbed by water vapor in the atmosphere diminishing the light of such wave lengths that can reach the telescope. At fourteen thousand feet, the telescope would be well above most atmospheric water vapor.

Funding of the magnitude needed from California seemed out of the question. Requests for federal funding would lay the project open to the politics of the astronomical community. In general, the National Science Foundation supports centralized facilities for the use of astronomers such as Kitt Peak and Cerro Telolo in Chile. However, the access to such facilities must of necessity be widely shared and thus infrequent. The Lick astronomers emphasized the need for some important studies to be assured of sustained time on the telescope—at least a few nights per month over a period of several months—which could not be had at the national observatories.

Early on in this project, given my long acquaintance with the Caltech astronomers, I had proposed that Caltech be brought into the project to share in the design and the cost. This idea was opposed by the Lick astronomers. They wished to retain full control of the project and they saw the telescope as the means to restore the preeminence of Lick over the Hale observatories. It should be added that the rather proprietary attitude of Caltech and the Carnegie Institute astronomers toward the Palomar facility and their long-standing reluctance to share its use had not endeared them to the astronomical community. At this time, in view of our growing funding problem, I again raised the possibility of a Caltech connection; however, the time was not yet ripe.

On 23 August 1983, Joe Calmes, the assistant to the director of Lick Observatory, received a telephone call from a Mr. Edward Kain in San Jose. Mr. Kain's accountant was Bill Unruh, an astronomy buff who had developed a great interest in the history of astronomy in general and Lick Observatory in particular, and who gave popular lectures on this subject at Lick to summer visitors. While doing Mr. Kain's taxes, Unruh had mentioned the plan for the new telescope and our inability thus far to find an "angel." Mr. Kain had been intrigued by this discussion and had a suggestion of a possible donor, Mrs. Marion Hoffman, Mr. Kain's sister who lived in Los Angeles. Her husband, Max Hoffman, had died two years earlier and Mrs. Hoffman was seeking a unique memorial for him.

We had earlier agreed that if a donor provided half or more of the construction cost (then estimated to be seventy-two million dollars) he or she could name the observatory. It was indicated, therefore, to Mr.

Kain that such a memorial would require a gift of at least thirty-six million dollars. He did not think that such an amount was infeasible. We investigated and found that Max Hoffman had likely left a substantial fortune acquired by his ownership of the import licenses for VW and BMW cars into the United States since 1946.

Through Mr. Kain, Mrs. Hoffman was contacted and the telescope project was explained to her. It developed that her late husband had had a lifelong interest in machinery and technological advancement. The idea of this unique, highly advanced, and novel telescope appealed to her as a potential memorial. The sum required seemed impressive but not daunting. She would have to consult with her accountants and lawyers.

Negotiations with Mrs. Hoffman proceeded rather slowly. We learned that she herself was not well and made repeated visits to a clinic in Cleveland. President Saxon had retired in 1983, and he was replaced by David Gardner, then president of the University of Utah. I had visited Gardner in Salt Lake City after his selection and had briefed him on the telescope project, in addition to the state of affairs at UC Santa Cruz. Although not a scientist, Gardner recognized the significance of the project and undertook to carry it forward.

Clearly at this stage the president's office had to be brought into the negotiations with Mrs. Hoffman. On 15 December 1983, President Gardner met with Mrs. Hoffman in Los Angeles. Agreement was reached on the size and nature of the gift (to be worth thirty-six million dollars) and the university's commitment with regards to the naming and securing the additional funding. Legal documents were to be drawn up. The next day, Mrs. Hoffman passed away from throat cancer.

The funds had earlier been transferred to a Hoffman Foundation. The three trustees of this foundation were Mrs. Hoffman, Ms. Ursula Niarakis, her longtime secretary, and Mrs. Doris Chaho, her sister. In the event of the death of any one trustee, the remaining two were to select a third. It soon developed that the two remaining trustees had no great affection for each other. They were completely unable to agree on the choice of a third trustee. It also developed that one of the trustees, while grudgingly conceding the intent of Mrs. Hoffman to fund the telescope, was less than enthused about this use of the funds. Matters ended up in probate court in New York and the university did in fact receive cash and title to tangibles worth thirty-six million dollars— the largest single gift in the history of the University of California.

We still needed to raise the additional thirty-six million dollars. In early 1984, I revived the notion of a partnership with Caltech as the other eminent astronomical center in California. The Lick astronomers were still not enthused, but with the prospect of a wondrous telescope looming, they acquiesced. President Gardner convened a meeting of the four chancellors, which brought general agreement. Chuck Young and I had a lunch meeting with Murph Goldberger, president of Caltech, to explore the idea. He was favorably inclined and agreed to discuss the idea with astronomers and others at Caltech. Further discussion proceeded through President Gardner's office. It was agreed that Caltech would undertake to raise twenty-five million dollars toward the telescope in return for which they would receive a proportionate share of the observing time.

It was in late August that I learned, via Harold Ticho, the UCLA representative on the executive committee, of a startling development. President Goldberger, in seeking to raise the twenty-five million dollars, had solicited several sources for gifts of five million dollars. He had received one such pledge when he received a telephone call from Mr. Howard B. Keck, a Caltech trustee. Mr. Keck's father, William M., had established the Superior Oil Company, which had been sold some years later for several hundreds of millions. A large share of the proceeds had gone to create the Keck Trust, which in turn supported the Keck Foundation of which Howard Keck was the chairman. Howard Keck proposed that the foundation provide, instead of five million dollars, the entire seventy-two million dollars to make this a Caltech telescope and to be named the William M. Keck Telescope. To Caltech's credit, it was not willing to take UC's design and build its own telescope, but the seventy-two million dollars was irresistible.

But what of the Hoffman estate? A rather ingenious proposal was formulated. Two ten-meter telescopes would be built, one to be named after Keck, the other after Hoffman. The cost for two would be considerably less than twice that of one. Both would be located on Mauna Kea about a hundred yards apart. They could be operated separately, providing twice the observing time, or jointly. In the infrared, they could be operated in an interferometric mode to provide extraordinary resolving power; in the visible region, their output could be pooled, providing no better resolution but twice the light-gathering power.

Ingenious, but unacceptable. The fractious and divided trustees of the Hoffman estate found this idea unacceptable. The telescope would no longer be a unique memorial and they seized on this opportunity to withdraw from the project. UC, with great regret, returned the largest gift it had ever received. The trustees divided the thirty-six million dollars to create two separate foundations, one for each with eighteen million dollars apiece.

Caltech and UC established a third corporation, CARA (California Astronomical Research Association), with a joint board of directors to oversee the construction of the telescope and its future operations. Caltech was to provide the construction funding, UC to provide operating funds for a period of years until its contribution equaled the construction cost. Thereafter, operating costs would be shared. Observing time would be shared equally and allocated by a joint committee. "First light" was seen in December 1990 and the telescope is now in full operation. Also, plans are underway, with a sizable second grant from the Keck Foundation, for the adjacent second telescope to form the "binocular."

What now would the two Hoffman Foundations do with their eighteen million dollars each? Could either be induced to devote the bulk of this money to a worthy technical or scientific project? The prospect coalesced several ideas in my head, answers to several questions that fit neatly into what has become the Human Genome Project.

As chancellor, I had become at least peripherally involved with several projects of Big Science. Indeed, harking back to Radiation Laboratory days, I had seen the power and progress made possible by the coordinated actions of diverse talents on a large scale, with ample funding. The ten-meter telescope was such a project, although on a relatively modest scale. Several of our astronomers were also involved in the Hubble Telescope project, an enterprise of five hundred million to one billion dollars. At chancellors' meetings we had had frequent discussions of the leading role of the university in the formulation of a proposal by the state to locate the proposed giant accelerator, the superconducting supercollider, a six billion dollar project, within California. Some of the UCSC faculty working in high-energy physics were directly involved.

Biology had no comparable Big Science projects. It was clear that other areas of science—physics, astronomy, space science—were not bashful about seeking relatively large sums of money to support their programs. Big Science, per se, was not a virtue. But were there important areas of biological research that were not being adequately explored because biologists were not thinking on an adequate scale?

The characteristic of Big Science projects in other fields was that they provided a facility that would be essential to further advance in the field. Biology did not seem to need a comparable facility. What biology needed, however, was a massive information base—a detailed knowledge of the genetic structure of several key organisms, including—for obvious reasons—man.

Mapping of genetic factors had been initiated with drosophila in the early years of the century. Extensive maps had also been developed since World War II for bacteria, yeast, maize, and mice. The first assignment of human genes to chromosomes other than the X became possible in the 1960s; by the mid-1980s, some four to five hundred genes had been so located. With the understanding that genes were composed of DNA and the development through recombinant DNA of the ability to isolate genes and sequence them, an extension of genetic mapping to the molecular level became possible. Some entire viral DNAs, ranging from a few thousand nucleotides to more than 100,000 had been sequenced. Efforts were underway to sequence entire bacterial DNAs (4.5 million base pairs) and the DNA of nematodes (about 80 million base pairs). Sequences of mouse or human DNA (genome 3×10^9 nucleotide pairs) were known only in certain small regions where they had been painstakingly worked out by groups interested in the genes of that region. Each group interested in a particular gene or gene cluster had to develop the appropriate technical expertise to sequence those genes.

What if a project could be undertaken to sequence, once and for all, the entire human genome. This would be Big Science, but the product would be an invaluable resource for all of biology and medicine. Was it feasible? What would it cost? How long would it take? How might it be organized?

Another question concerned the future of UC Santa Cruz. In its early stages, the campus had understandably attempted to develop a competence in the basic fields and disciplines expected of a university. But now with prospects of future growth brightened, the campus had to develop several centers of excellence, to focus a portion of the additional resources that would come with growth in the establishment of selected programs that would merit national and international attention. Naturally, I wanted to see such a center of excellence in biology. Given its significance, an institute for research on the human genome would accomplish this objective. Such an institute would be a natural component or a center for the human genome initiative and would insure that UCSC biology would be in the forefront position for many decades.

The establishment of such an institute, however, would require substantial funding. To be on a useful scale, it would require, I estimated, a building, equipment, and endowments adding up to perhaps twenty-five million dollars. The federal government was a possible source, but I knew that however attractive the concept, NIH or NSF would have to award any sum of this magnitude on a competitive basis, and in such a competition UCSC would never win out against Stanford, MIT, Caltech, Berkeley, and other "heavy hitters." If we had an institute already established, however, the research proposals would surely command support for specific projects from NIH or NSF. But we would need to raise the initial funding from private sources.

The eighteen million dollars now located in each of the Hoffman Foundations beckoned. The boldness and significance of the concept might appeal to one of the trustees. To avoid an unseemly scramble, President Gardner had decreed that any approaches to the Hoffman Foundation must pass through his office. Therefore, on 19 November 1984, I wrote him as follows:

Dear David:

Let me expand a bit on our brief discussion at the regents' meeting on Friday.

If the "Hoffmans" firmly intend to withdraw from the TMT project, then I have another project that we might propose to them. It is an opportunity to play a major role in a historically unique event—the sequencing of the human genome.

A genome is the complete set of DNA instructions for the making of a species. The human genome is the complete set of instructions for a human being. We know that the haploid human genome is composed of some three billion nucleotide pairs (3×10^9). A few months ago, I posed to our biologists the question, Could the human genome now be sequenced, with extant technique, and in a reasonable time (in years)? If so, what scale of effort would be required? (Obviously, I had made a guess as to the answer.)

Their reply is enclosed. It can be done. We would need a building in which to house the Institute formed to carry out the project (cost approximately twenty-five million dollars), and we would need an operating budget of some five million dollars per year (in current dollars). Not at all extraordinary.

Clearly, the human genome will be sequenced. It will be done, once and for all time, providing a permanent and priceless addition to our knowledge.

In addition to satisfying our scientific curiosity, this knowledge will provide deep insight into other questions of interest. It will have major medical implications: we know that literally thousands of human ailments have genetic bases, in whole or part.

This knowledge will also have highly significant evolutionary implications. The biological differences between *Homo sapiens* and the chimpanzee are certainly due to the changes and rearrangements in the genomes of each as they have diverged from that of our common ancestor. To understand these changes will surely illuminate the ancient human quest to know what we are and where we came from.

The enterprise could be known as the Hoffman Project and, of course, the building could be named the Hoffman Laboratory or Institute. If we had the building and equipment, I feel quite confident we could obtain the operating funds from government and/or private sources.

Needless to say, should the "Hoffmans" not be interested in this project, I will intend to look elsewhere for funding.

I expected David Gardner, too, would be seized by the deep significance of the proposal, but then I'm a biologist and he is not. He wasn't seized. It must have seemed like another in a set of meritorious, costly proposals he was receiving from various parties on various campuses. After a few months it became apparent to me that funds from a Hoffman Foundation were not likely to be forthcoming, if indeed they were ever to be solicited. I would have to look elsewhere. Other sources came to mind—Arnold Beckman, David Packard, Gordon Getty, the Howard Hughes Foundation. Any of these *could* fund the project, but before I could make such approaches, I needed more certain ground as to the feasibility of the concept.

Until now, I had germinated the ideas in private and in discussion with some of the UCSC biologists—Harry Noller, Bob Edgar, Bob Ludwig. At first dubious, they too had soon been seized with the notion. We decided to convene a small workshop to explore the idea with the people who were most prominent in the sequencing field. As chancellor, I allocated the needed funds and we set the dates for 24–26 May 1985. The membership of the workshop included representatives of the most active sequencing groups, researchers interested in the development of associated automated instrumentation and computer techniques.

The initial mood at the workshop was clearly one of great skepticism about the project. As the various aspects were discussed, it became clear that many elements of the task were indeed feasible and that plausible advances in automation could well make practical the entire sequencing project. When thus pushed to confront the possible reality of such an enterprise, differences of opinion emerged as to its desirability. These were not resolved, but the workshop had changed the question in the minds of the participants from one of feasibility to one of desirability. From can to should.

I prepared the notes and conclusion report, which were sent to the participants and other interested parties. As I said in my letter of 5 June 1985 to Dr. Donald Fredrickson, the president of the Howard Hughes Institute:

Briefly the conclusions were:

(1) A *genetic map* of the human chromosomes providing well-defined markers (polymorphisms) at reasonable spacings along all of the chromosomes, to use as reference points, could be developed (in collaboration with outside groups) by a staff of perhaps twenty people in a two- to four-year period.

(2) A *physical map* of the human chromosomes providing a linearly ordered set of cosmid-size (thirty to forty thousand bases) DNA fragments could similarly be developed by a group of twenty people in two to four years.

(3) A complete *nucleotide sequence map* of the human chromosomes is not presently feasible with reasonable effort. Sequencing a few percent of the genome around selected markers and in carefully chosen regions *is* feasible, with a group of some thirty people working over a ten-year period. The availability of such sequences would undoubtedly be of great value. At the same time, it is quite reasonable to anticipate advances in and automation of sequencing technology such that the sequencing of the next few percent could be done with one fifth or one tenth of the man-years effort.

(4) There was general agreement that a *centralized effort* correlating genetic, physical, and sequence mapping, promoting the development of improved technologies, and actively fostering the application of this knowledge and approach to specific problems in human genetics, development, and physiology would be of great value.

Over the next year, I sought unsuccessfully to interest potential donors in support of this project. Somewhat näively, I believed that the project, to determine once and for all the genetic basis of man with all its rich and incalculable consequence, would surely seize the imagination of anyone with even a rudimentary scientific bent. Curiously, it did not. I also believe that the proposal would have been given greater credence and a better hearing had it been put forth by a more prominent, established institution—a Caltech, a Stanford, a Harvard. UCSC was an undistinguished spot on the map of biological research. Ideas should be evaluated purely on their merit, but in the real world, in the battle for attention and credence, that seldom happens.

Elsewhere, however, the concept was demonstrating that it was an idea whose time had come. My summary of the workshop was circulated in the biological community. The idea had come to the attention of the officials of a seemingly unlikely sponsor, the Department of Energy, and in particular to Charles DeLisi.

In fact, this was not so unlikely. The Department of Energy is accustomed to the management of Big Science and the DOE had in existence a biology program dating back to the Manhattan Project. From its inception, this program had been concerned with understanding the biological effects of radiation, but in recent years scientists in the program had moved opportunistically into related fields.

Using the availability of exceptionally powerful computing facilities, biologists at Los Alamos had established Genbank, the national repository for the DNA sequences that were emerging in a growing stream from biology laboratories. Approximately ten million nucleotides of sequence, from various DNA sources, were already on file at Genbank. At Livermore National Laboratory, biologists taking advantage of the local expertise in laser technology had applied this tool to the fractionation of chromosomes, modifying a biotechnique originally developed for cell sorting.

Thus, in the spring of 1986 DOE convened a conference to discuss the merits of a national program to sequence the human genome. Regrettably, a conflict made it impossible for me to attend. A second conference was held at Santa Fe in January 1987 on techniques for the automation of processes related to DNA sequencing. A DOE subcommittee, headed by Professor Ignacio Tinoco of Berkeley, was established in 1986 to provide advice as to the feasibility of the human genome sequencing project and the outline of a plan as how best to proceed. I served on that committee.

We recommended that such a program be established and that the initial emphasis be on the establishment of genetic and "physical" maps—sequentially ordered collections of ten to twenty kilobase tracts for each chromosome—while the development of automated machines for the more laborious task of nucleotide sequencing proceeded. At the same time, development of computer programs for management of the vast amount of data to be gathered would proceed. We recommended an initial budget of forty million dollars a year to ramp up to some two hundred million dollars per year over five years. We projected that a further ten-year expenditure at that rate would complete the program and provide the sequence. We also proposed that some definite goals be set such as the sequence of one or two of the smaller human chromosomes by a definite date.

As this program moved forward, voices of opposition began to be heard. One faction argued against the introduction of Big Science into biology especially if it were to come at the expense of the current smaller-scale projects. The objection was in part self-interest, in part a conviction of the superiority of investigator-driven initiative, and in part a simple unfamiliarity with and fear of Big Science. A more glib objection raised was that much of the sequence information would be useless—that only a few percent of the human genome was meaningful and the great bulk was "garbage." Given our ignorance of so much of the genome, the evidence to support this position was limited; further, one person's garbage can be another's treasure. The potential utility of so-called "nonsense" DNA, for evolutionary or anthropological studies, or its role in control processes, is as yet quite unknown.

Still another objection related to the emphasis on human DNA. Admittedly, for the clarification of the present issues in biology, the complete sequence of *E. coli* DNA, drosophila DNA, or mouse DNA might be more useful, but to obtain a commitment of perhaps three billion dollars, the sequence of human DNA with its potential for medical insight was far more justifiable. And once the apparatus for the determination of the human DNA sequence was established, its application to the other genomes of interest would not be difficult.

The idea had an inherent natural appeal that has in fact swept all before it. The project is now underway. I merely provided the push to start the snowball rolling. It did not benefit UCSC as I had hoped, but it will surely benefit humanity and I am pleased about that. To have come from the nearly total ignorance of cellular substructure and molecular process in the 1930s to the prospect of a total knowledge of the human genome in the 1990s—what a remarkable journey of discovery.

The memorial that Mrs. Hoffman wanted to create for her husband was never realized, but from that thought so much has come. Perhaps it is sometimes truly "the thought that counts."

30

Transition 6

When the parents of the freshmen look too young to
have college-age children, it is time to move on.

Retirement from the chancellorship was mandatory at age sixty-seven,
but it was time to go. Ten years of administration were enough. The
routine of the academic year—meeting the new freshmen, reporting to
the academic senate, the fall UCSC foundation meeting, the staff
Christmas party at University House, the monthly council of chancellors meetings, the monthly regents' meetings, the regular senate meetings, the spring foundation meeting, the myriad annual receptions for
Friends of ———, the ARCS scholars' lunch, the athletic awards party,
the staff awards picnic, the multiple commencements, the end-of-the-
year faculty party, the annual budget sessions, all punctuated by student
protests over some issue or other—had indeed become routine and
begun to pall.

Old issues, never settled, only patched, began to recur—affirmative
action, financial aid, child care, ethnic studies, community unhappiness
with campus growth, yet another provost for Kresge College, yet another dean for natural sciences, yet another furor in the music board.
The freshmen seemed younger every year. The freshmen in 1986 had
been born around 1968. To them the Kennedy assassination, the Vietnam War, and Watergate were as much ancient history as were World

War II, the Great Depression, the American Revolution, and the Roman Empire.

Repetition breeds cynicism. Hearing the same time-worn complaints from each new crop of students, watching and bearing their patterned protests and antics, it became harder each year to muster an open mind, to remember that it was new to them, to respect their need for growth and expression. I had reviewed the arguments many times and come by then to firm conclusions. It thus became hard to enter into a true dialogue with students, to present a paternal rather than a dictatorial face. Having heard all of the arguments many times over, my mind tended to wander, to concentrate on the student personalities in lieu of the substance—callow but earnest, näive but impassioned, incredibly self-centered, incredibly arrogant, and convinced at their tender age of their superior wisdom.

My natural tendency is toward a "liberal" orientation. A liberal philosophy necessarily is based on a belief that people are basically good and well motivated and will, if left free, act for the common weal. Contact with succeeding generations of students, however, corrodes that belief and reminds one how much education is needed to achieve even a modest maturity. An ability to view this frothy scene with a sense of humor had been a saving grace, but even this had begun to ebb.

To be a chancellor is a baptism in a whirlpool spun ever faster by the conflicting forces of modern society, for these forces correctly focus on the university as the fountainhead of the future—and seek to influence its direction by acting on its titular head. But the chancellor cannot directly oblige any of them for, in fact, he or she has very limited power to alter its course. Who does? In truth, very often, no one. Inertia dominates.

~~~~~~~~

UC Santa Cruz was one of a group of colleges and universities launched in the 1960s (e.g., Evergreen, Hampshire, Old Westbury, New College) to "reform" higher education. All fell on hard times. Santa Cruz survived only by virtue of its lifeline to the UC system.

The founders of UC Santa Cruz knew more what they were *against* (disciplinary and research emphasis with resultant faculty indifference to undergraduate education) than what they were *for* and how to accomplish it. Incompletely conceived, the birth of the campus was premature. The infant university was then afflicted with the general campus turmoil of the 1960s and the accompanying drastic changes in student attitudes, while its congenital deficiencies grew increasingly salient.

To create a campus designed in contradiction to the role of UC as set forth in the master plan for higher education in California was quixotic at best. The conflicting objectives of colleges and disciplines charged every decision with resultant partisan bitterness: if one favored the colleges, it was "lowering standards"; if one favored the disciplines, it was "a betrayal of the campus vision." In short, it was a no-win situation.

Does higher education need reform? In my view, yes. Apart from the claims of myriad special interest groups, American higher education in general succeeds admirably in the provision of disciplinary education (indeed, its appeal is international) but largely fails in the provision of broader inter- and cross-disciplinary perspectives and concepts. There is no consensus on the nature of a "liberal" education for the twenty-first century and little incentive to develop one. Lacking this, American universities often become, in good part, training grounds for future white-collar workers and holding tanks for youth in prolonged adolescence.

If reform is to come anywhere, it will require a conception that is thought through well to reflect external context; faculty motivations, incentives, and availability; and student motivations and interests, as well as more abstract educational goals. The thorniest problems involve curriculum and faculty responsibilities. These issues must be thought through *before* the university is launched, before its energies are consumed in the construction of facilities, the organization of classes, and the stewardship of students. Several years, or at least several summers, of sustained thought and discussion by a committed group of faculty would be needed to develop a consistent and coherent curriculum and work out the details of its implementation before it is attempted.

The new enterprise must be assured of resources adequate to the tasks. The concept should embody a clear vision, but promises should not be made nor expectations raised that cannot be fulfilled and lead only to continuing frustration. Older standards of excellence should not be abandoned without explicit—and tentative—justification. An inherent element of any reform program should be periodic evaluation of its successes and failures and a willingness to "reform" the reforms. The reform movement should not succumb to the pathology of "instant tradition," in which early casual actions become unalterable sacred precedent. And the conductors of educational experiments should at all times have a conscience with respect to the students whose lives and future careers are in their hands and to the junior faculty whose careers are precious and precarious and must similarly be nurtured.

~~~~~~~~~

There were the accustomed bittersweet farewell dinners and events. Many people seemed genuinely sorry to see us leave and I was deeply touched. Karen was named "Woman of the Year" by the Santa Cruz Chamber of Commerce for her many community activities—the first time anyone from the campus had ever been so recognized in the community. A faculty chair was endowed in my name, thanks in large measure to my old college friend Art Graham and his wife Carol. Even faculty and community people with whom I had been much at odds seemed to mellow and sheathe their swords and wish us well as the time drew near.

But, it was time to go. Science beckoned.

Ending Well

31

Return to Science

"You can't go home again" was never truer. It had been ten years since I had, in any sense, been a practicing scientist, ten years of revolution in my very own field of molecular biology. The advent of recombinant DNA and cloning, combined with DNA sequencing, polynucleotide synthesis, and genetic engineering, had broken a long-standing research barrier by permitting the isolation and identification and modification of individual genes from higher organisms. Whole fields of study—some long stagnant, some brand-new—had opened up in developmental biology, cell biology, virology and microbiology, neurobiology, and nearly every aspect of medicine.

In 1977, just prior to becoming chancellor, I had written a review for *Annual Reviews of Biochemistry* on the then new topic of recombinant DNA. After ten years of sophisticated genetic engineering of recombinant DNA vectors and the development of advanced techniques and instrumentation, that review was wholly obsolete. I had read occasional semi-popular articles in *Science* or *Scientific American,* but I had not kept up at all with the technical advances—which, in any case, would be meaningless without actual experience in their application.

Even more profound than the advances in technique were the enlarged perspectives and the deeper insights now available. Biologists have long marveled at the variety of living forms in the world of nature around us. We are now learning that the variety of biochemical mechanisms and processes within these organisms, built on a base of ancient, common strategies and structures, is even more exuberant, even more extraordinary. Over billions of years, in myriads of species, nature the experimenter has explored and exploited all the changes, the combinations, the cyclic sequences of events to produce programs—cellular dramas—with common themes but of dazzling intricacy and variety.

If I were to reenter science, I needed to travel the usual entry track—as, in effect, a postdoctoral fellow. Caltech seemed the obvious place to do this. I knew most of the people; I knew the structures, physical and organizational; I knew the invaluable ambience. After ten years at UC, I was entitled to a year of sabbatical leave, so I would not need a salary. Caltech was willing to provide me with a small basement office and, most important, access for the year.

However, where would I go after Caltech? My faculty appointment was at Santa Cruz. In light of my own experience with a former chancellor on campus, I felt it would be much preferable not to be a faculty member at the same institution where I had been chancellor. It is awkward for the new chancellor to find a predecessor "popping up," in a sense looking over his or her shoulder. It would be difficult to avoid becoming involved in campus issues in which I had invested so much thought and energy, and I would indeed be pressured to do so by one faction or another. But I definitely should not. They are now the new chancellor's problems, opportunities, and responsibilities.

UC Santa Barbara appealed. It was a young and developing campus, with a growing reputation in the sciences, especially physical science. It had some distinguished biologists whom I knew. The climate was benign. And I had a house there, which I had built in the mid-1970s to use for vacations and which had been rented out all of the Santa Cruz years. The Santa Barbara biologists seemed pleased to have me join them. My appointment would remain at Santa Cruz; I would be on "temporary" assignment to Santa Barbara until my retirement as a faculty member, which would be mandatory in three years.

We rented an older house in Pasadena, near Caltech, and arrived in the midst of a heat wave, just in time for the jolting aftershock of the Whittier earthquake. But Pasadena was still familiar.

Returning to science after an enforced abstinence was a joyous journey. I could savor anew even the process as well as the content: the familiar rhythms and flow of a scientific meeting, the large general lectures and the smaller specialized sessions, the little knots of people, some old friends, some new acolytes, discussing a presentation—or where best to go for dinner—and above all the pervasive sense of progress, of advance, since the prior meeting.

Even the easy rituals of the late afternoon seminars were fresh and pleasing: the faculty largely in the first few rows, the students scattered behind them; the introduction of the speaker, the standard acknowledgments to coworkers and students before launching into substance, the background, the recent results, the future plans; then the discussion period, sometimes respectful, sometimes probing, sometimes brash, sometimes almost harsh, but impersonal. All the familiar and heuristic patterns, time-tested and leading erratically but surely to truth, once again shone and refreshed.

Much of the fall quarter was spent reading, talking to everyone in biology, attending numerous seminars, and catching up on the state of the science. But I clearly needed to get into the laboratory. With recombinant DNA, developmental biology had emerged from a fifty-year slumber. Finally, one could begin to study the genetic and biochemical components, the hierarchy of controls that underlay the intricate processes leading from a single fertilized egg cell to a mature organism. Eric Davidson's laboratory was most active in this field, so I signed on with him as an apprentice postdoctoral fellow.

The postdoctoral fellow in science has in many ways an idyllic existence. He or she has no responsibility other than to come in and perform research. The professor provides the laboratory, the facilities, the funds for supplies. Undistracted by teaching, committees, or fund-raising and with such recent training, the fellow is ideally positioned to concentrate for a few years on problems at the most advanced edge of the field. Of course, I had one advantage over the other postdoctoral fellows—I did not have to worry about finding a permanent position in two or three years.

Getting back to the laboratory bench, doing experiments with my own hands, coming in in the morning to see how an overnight experiment had worked out, reviewing results, and exchanging ideas with others in the laboratory was sheer pleasure. Initially, I was quite ignorant of the new techniques, but other fellows and students were most helpful. Once past the early blunders, most of the techniques are actually rather simple to use, save for the occasional total failure due to a mental lapse or, more often, a bad batch of reagents or an equipment malfunction.

Development implies a programmed pattern of controlled differential gene expression in the varied cells of the developing embryo. The research in Davidson's laboratory used embryos of sea urchins—a classical object of embryonic research—which can be obtained in large numbers and accurately synchronized. The object of the research at that time was to isolate and identify factors—proteins—responsible for turning on and off specific genes at specific stages of development, and then locate the genes responsible for these factors. Ultimately, we hoped to arrange all of these in an hierarchical structure, beginning with the fertilized egg, that could explain the appearance of different functions and organs at the different stages of development. Some genes controlling analogous factors in drosophila had been isolated and sequenced. It was plausible that similar genes existed in sea urchins. If so, techniques existed to identify, and then isolate, these by their homology to the drosophila genes. I set out to do this.

In the laboratory, I was immediately struck by the ready commercial availability of sophisticated biological reagents. A whole ancillary industry had sprung up to provide the many enzymes, polynucleotides, engineered plasmids, viruses, cell lines, and highly purified reagents needed for modern molecular biology. I recalled the era when one had to go to the stockyards to obtain intestines to prepare alkaline phosphatase and thymus glands for DNA. As recently as the 1970s, we had had to prepare our own restriction enzymes. Now a hundred different varieties could be purchased at quite reasonable cost. This development greatly accelerates the pace of research.

Still, as always, the research went more slowly than I hoped, but by summer I had had some success. Now it was time to move to Santa Barbara, where it would be difficult for me to continue this project on my own. But I had mastered a variety of essential techniques.

The biology program at Santa Barbara, because of leaves of absence and pending retirements, needed me to perform some of their essential teaching for a few years. I welcomed the challenge. And it really was a challenge.

The third quarter—thirty lectures—of the required biochemistry course concerned nucleic acids and protein synthesis. This had been my field of interest and research for decades and I had taught varied aspects of this subject at Caltech.

As I prepared to teach this course after a ten-year hiatus, I was astonished and thrilled by the progress made in the past decade. In every sector, the advances had been simply extraordinary: in the intricacies and nuances of DNA structure (no longer the simple rigid double helix); in the still deepening complexities of DNA replication (initiation, strand elongation, termination); in the much expanded understanding of DNA transcription and its control; in the knowledge of the processing of RNA transcripts (capping, splicing, tailing); in the comprehension of amino acid activation and ribosomal structure and the many facets of protein synthesis (initiation, elongation, termination); in the variety and importance of DNA repair processes; in the growing knowledge of chromosomal structure and the enzymology of genetic recombination; and, of course, in the whole world of recombinant DNA with its skillfully designed vectors, the elegant methods of oligonucleotide synthesis and DNA sequencing, the wealth of restriction enzymes, the new polymerase chain reaction, and so on—a true biological engineering.

Practically, I had difficulty in planning the time required to treat each topic, and in the end I estimated that over half of the material I taught in the course had simply not been known ten years earlier.

Between organizing the lectures, preparing transparencies for projection and Xeroxed handouts for eighty students, meeting with students who had questions, planning the discussion sections with the teaching assistant who would lead them, and giving and grading midterm and final examinations, the course was a ten-week marathon occupying nearly every waking hour. I learned a lot, and it was truly enjoyable to be able to talk about science and once again interact with students in a nonconfrontational mode.

A series of ten lectures on the cell nucleus in a graduate course in cell biology was a similar, but briefer experience. Quite different, however, was a series of lectures on genetics to a beginning biology course for students who did not plan to major in biology. A popular way to satisfy the science portion of the general education requirement, the course was limited to 475 students only because of the size of the lecture room. The room was actually a concert hall with no blackboard and a sea of faces. In such a course there can be very little direct interaction with students. The presentation must be very largely visual. I used slides, transparencies, films, videotapes (e.g., *Nova* programs)—whatever I could find that was appropriate. Today's TV-habituated students seem better able to learn from visual presentation than from textbooks. But as they are accustomed to the high technical standards of commercial television, the visual material has to be of a similar quality.

Soon after I arrived at Santa Barbara, I learned of the research of Professor Paul Hansma in the physics department. Hansma had developed an "atomic force microscope" with which he had been able, with favorable substrates, to obtain atomic resolution on surfaces. It occurred to me that this instrument might, just might, be able to resolve the individual bases in DNA and if so permit one to achieve the direct sequencing of a DNA chain. Such an accomplishment would, of course, greatly advance the Human Genome Project. This was and is a "long shot," a "high-risk" experiment with the possibility of a dramatic result. It was not an experiment that one would give to a graduate student who must complete a thesis and write some papers for his career, but I could take the chance.

Paul Hansma was intrigued with this possibility and together with his wife Helen, who is a biologist, we set out. We have been looking at simple polynucleotides of known sequence. In such research there are always many possible approaches (Should the nucleic acid be dry, under water, under a nonaqueous solvent?) and technical difficulties (What supporting surface should be used? How can we keep the nucleic acid from moving about? Is temperature important? Can we minimize possible damage to the nucleic acid or is that not a problem?) To date we have had modest success. Our resolution has been such that, under favorable conditions, we can detect the presence of the individual nucleotides in a chain but cannot differentiate them. Could we modify them to make them more distinctive? Would other modes of mounting the DNA be more productive?

The research continues, and it is fun. And the reports of progress in each new journal, the new discoveries presented in the weekly seminars, and the new insights that come from reflecting on these all stir the same excitement, the same pleasure and wonder, as always. Science is a glory of the human mind and these are glorious times in which to be a biologist.

32

"We Happy Few"

I am a scientist, a member of a most fortunate species. The lives of most people are filled with ephemera. All too soon, much of humanity becomes mired in the tepid tracks of their short lives. But a happy few of us have the privilege to live with and explore the eternal, to feel the wind at the ever-advancing edge of human knowledge, and to peer into and progressively reveal the dim shapes of the unknown. Scientists find the natural world endlessly absorbing, a fertile obsession. In the exploration of nature, we can exercise our full imaginations, our most acute logic, our deepest curiosity—and the knowledge we gain will endure throughout time.

In recent years, some humanists have sought to blur the distinction between science and other more ephemeral forms of human activity, to distance science from the "eternal." The "deconstructionists" have claimed that scientific truth is only an outcome of "negotiations" between scientists in the laboratory, that the "natural world" emanates from the "social," and that, as in other human spheres, "truth" in science is not independent of the exercise of power. While such views have gained credence in some circles, they are painfully superficial and do not correspond to the reality of the world of science I know. Science does differ from other human pursuits in that there is an *external* standard of truth that must be met. There are basic properties of matter and life independent of human intervention. Which areas of nature are explored is, to be sure, socially determined; the framework and language in which the laws and properties of nature are described are of course human constructs. But there is an external reality, an external truth that cannot be falsified, that is independent of the murky claims of human motivation, and that is as close to "eternal" as humans may approach.

〰〰〰

"The glass is half full; the glass is half empty." One can view life either way. I believe you have a choice. In an interview at age ninety, the great dancer Martha Graham suggested that every choice is a sacrifice of the road not taken. But, alternatively, every choice is a cause for celebration—celebration that we of all creatures have the capacity for choice. We are not limited to instinctual responses but can use forethought and will to guide our actions. In all of the known universe, we, *Homo sapiens,* are the wild card. In all else of nature, the past creates the present, the present creates the future. What is, is because what was, was. But, with us time closes back on itself. The envisioned future affects the present. And science most powerfully permits us to understand the past and the present and to envision the future.

〰〰〰

We are all limited human beings with only special innate talents, with only finite perspectives, with depths unplumbed, heights unscaled, and connections never made. We scientists are a type, a breed. We are oriented to things not people, to clean abstractions not messy realities, to reason more than emotion. We are inexhaustibly curious about *how* things work, but not *why,* for motive is only real in the people world we eschew.

We scientists inhabit a distinctive culture. We live in both a narrow time frame and a very very long time frame. In our daily work we focus on the near past, the present, and the near future; in our science we range from the beginning to the end of time. But human history is far less real to us. The rapid progress of science renders our past much more distant; unlike many groups, we do not harbor ancient grudges for historic wrongs.

Advances in science are permanent. The knowledge, once gained, is forever ours. In contrast, advances in social affairs, however hard won, seem fragile and temporal—always at risk of the struggle resuming, reversing, or transferring to another arena.

Unlike anthropologists or economists, authors or poets, theologians or politicians, we natural scientists have the luxury of a single truth.

There is only one proton mass, one periodic table, one genetic code. In consequence, science, during my career, has been essentially egalitarian. Nature is the only source of ultimate authority. Before nature, a world outside of man, a reality independent of human design or desire, we are all equal. Nature is not deceitful and nature does not play tricks. We scientists may deceive ourselves, as by our all-too-human quest for a superficial simplicity, and thus for a time we may overlook a deeper truth. But, as research continues, nature alerts us to our myopia.

Critics and revisionists may now argue that *access* to nature is becoming less egalitarian in this emerging era of intensive instrumentation. We should be consciously concerned that such barriers be minimized.

The purpose of science is to create an inner world that predictably matches the external world. Others create inner worlds for varied purposes, constructive or frivolous, more ordered or more chaotic, more spontaneous or more reflexive. Science requires imagination on which it then imposes discipline. The knowledge that there is an external truth is both a comfort and a trial for the scientist. Unlike the artist who must rely on either an inner voice or an uncertain outer acclaim to validate his work, the scientist knows there is a sure single objective judgment that will be rendered on his experiments: Have they guided us to an accurate perception of a reality? The scientist is therefore more constrained and potentially more vexed. The real world is there waiting to be discovered—if one can find the path—but unlike the artist, the scientist cannot be content with his own "truth." Even imagination and elegance are of little merit if in the end they describe only an unreal world.

One of the values of a life in science is that one does not have to cope—at least not very often—with venal or corrupt people. Scientists are not paragons—ego, vanity, greed are scarcely unknown to us—but out-and-out corruption, lies, deceit, thievery, abuse of power, and bigotry are rare. The discipline and earnestness of the enterprise seem to make such acts too petty, too unworthy. Or perhaps, at least in biology, the true nuggets have been so plentiful and so close at hand as to curb desire to steal from an associate or to gold-plate a common pebble.

Beyond the hours devoted to survival—a need that our society now fortunately satisfies for almost all—how does one live the other hours? I have been fortunate to find true pleasure and excitement in this activity of science, which I can regard as wholesome, deeply meaningful, and enduring. MIT transformed me into a goal-oriented person and equipped me with the means to set and achieve goals. Because I have thus been goal-oriented, I have had scant patience with those individuals who constantly look back to a past episode—their war experiences, athletic performances, or student "activism"—as the high points of their lives. As a participant in science, one can always believe with confidence that the best is yet to come. For, most remarkably, nature—even biology with its intricacy developed by three billion years of evolution—seems indefinitely penetrable to human reason.

~~~~~~~

I was fortunate enough to enter a field of transcendent importance—nucleic acid research—at the time of its infancy, when one could actually know everything there was to know about that field, be acquainted with all of the major contributors to the field, and read all of the current literature. Today this would be impossible.

In retrospect, I see that I have had a penchant to choose nonmainstream courses. To minimize competition? Perhaps. In hope that virgin territory would have a greater chance of a rich, undiscovered lode? Perhaps. Or simply because unexplored regions gave one greater freedom to chart one's own path? Perhaps. At each major transition, I have made such a choice. Biophysics was such a choice, as were DNA, ϕX, and UC Santa Cruz. All proved interesting; most proved felicitous.

The advances in biology since I entered the field were at that time well nigh inconceivable. In the late 1930s, we were nibbling at the edges of the great problems—heredity, development, homeostasis, brain function—grasping at straws to construct ambitious theories in the absence of evidence. Today we are attacking head-on the core of these questions. These advances have brought remarkable insights. Especially, those in our understanding of heredity have been the most profound, the most seminal.

We have long recognized that one of the distinctive features of *Homo sapiens* is that we, among all the species on earth, are self-aware—aware of our individuality, aware of the mystery of our origin, aware of the future and of our ultimate mortality. Today we are increasingly self-aware in a very different sense, for we are now becoming aware of the intricate machinery within us. No other species knows that it has a circulatory system, a hormonal system, a nervous system. And now we are deepening that awareness to the genetic level, unraveling and exposing the chemical machinery—the complex array of genes—that through controlled expression, repression, and multiple interactions brings about our growth, differentiation, and maturation and produces (miraculously) the very intelligence that permits us to decipher this extraordinary process. How remarkable!

In the deepest sense, we are who we are because of our genes. Genes provide our physical framework, much of the specific basis for personality, and the raw material for intellect. (Circumstance, environment, and culture map the specific routes for intellect.) If we are ever to find out who we are and how, via evolution, we came to be who we are, we need to understand our genes, our biological inheritance, in detail. When we have it, what then will we do with this knowledge? We cannot escape this question, which is both enthralling and chilling. A species potentially able to plot its own genetic destiny is truly unprecedented. What wisdom can we summon to guide such a venture? Are there principles of increment, rules, or uncertainty, principles to govern manageable rates or size of genetic change?

Ten thousand years of cultural evolution have brought us to this point—unplanned, uneven, the turbulent but continuous history of our species. Much of the continuity has been provided by the continuity of our biological nature—our physical needs, our psychological and intellectual range, our limited life span. Were these to change, what consequences would flow? Such questions, aborning in the laboratories of our time, would seem to beggar the political and socioeconomic concerns that dominate our news.

From the time of the invention of writing, men have sought for the hidden tablet or papyrus on which would be inscribed the reason for our existence in this world, on this planet in this star-lit universe. How poetic that we now find the key inscribed in the nucleus of every cell of our body. Here in our genome is written in DNA letters the history, the evolution of our species over billions of years. The message is faded in places, tattered by the insults of the eons, but of necessity valid and functional in its vital parts. When Galileo discovered that he could describe the motions of objects with simple mathematical formulas, he felt that he had discovered the language in which God created the universe. Today we might say that we have discovered the language in which God created life.

~~~~~~

A life in science, a life on the edge of knowledge, can provide endless fascination and intellectual reward. There one finds a continuing sense of progress, of challenges met and challenges overcome. The scientific problems of my youth have been resolved into a progression of deeper and deeper questions. Most still challenge us, albeit in different terms. To some we have found definitive solutions, but even these often lead on to further, previously unimagined questions.

To know science is to see a wondrous pageant enacted over centuries. The curtain rises, at first slowly, on a scene of mists and strange, indistinct shapes amid the half-light. The protagonist is humankind, a stranger in this strange world, a part of it yet apart from it—puzzled, filled with wonder as to where he is, how he came to be here, what guides this cosmos, and what his destiny is. Gradually, here and there, the mists dissolve, some shapes become clear, some mysteries recede, and some vanish in the clear light of knowledge. Yet others emerge in their wake, formerly hidden in the background. A pageant incomplete, unfolding over the generations.

Until recently, scientists could also take satisfaction in the thought that their activities were an unalloyed good—that knowledge was good and more knowledge better. Human intelligence has been the principal agent of human progress. And science, the disciplined search for new knowledge, has, at least in recent times, formed the leading edge of intellectual advance. Yet strangely, in our time, as scientific knowledge has advanced in every field, as the powers of applied science have multiplied again and again, the public perception of science has markedly darkened. Science, once the Promethean reliever of toil, the bringer of light and health, the transcender of space and time, the enabler of so many human dreams, has become science the handmaiden of pollution, the abettor of overpopulation, the fountain of apocalyptic military technologies.

Science is a mutator gene in our society. Could it mutate itself out of existence by mutating the society to become resistant to its product—resistant and fearful of innovation, or so comfortable that few will choose to enter such a demanding career? Some portents are already clear. Many futuristic novels such as *The Time Machine* by H. G. Wells and *Brave New World* by Aldous Huxley envision a world with two distinct classes of people—a technologically advanced, dominant class and a more primitive, naturalistic class living a simple but hazardous life.

Once I thought that these scenarios merely reflected British class distinctions. But one sees the same bifurcation occurring in the United States, between a scientifically literate, technology-prone culture and an antitechnology, back-to-nature cult opposed to "the rape of the earth," "research on animals," and "technological enslavement." The technical illiteracy of most of our people has led to a growing alienation from the "incomprehensible" manmade world and a growing passion to return to the romanticized, intuitively grasped world of nature (to be sure, as tamed and made accessible by technology). From our beginnings, humans have sought to leave an imprint on the earth, an evidence of their existence—as cave drawings or pyramids, as statues or portraits, as castles or factories. Today, for the first time, we have a generation that seeks to remove human imprints, to return the earth to a more primitive state.

Most scientists recoil from the evidence of these perceptions that their work is not universally admired. They respond with the argument that their role is to discover knowledge, that they are not responsible for its use or abuse. Partly, this is true. Indeed, one can hardly foresee the longer-range uses of a basic discovery. When Einstein formulated his famous equation relating matter and energy, he could hardly have foreseen that it would find application in an atomic weapon. But as research has become more and more rapidly the source of industrial development, that excuse has become an evasion. Such evasion is clearly evident in the approach of scientists to the setting of priorities within science. The issue tends quickly to gravitate to the lower brain centers, numbing the mind and inflaming the emotions.

Scientists believe they serve, in their discipline, an inherently noble cause. They believe that a multitude of minds including their own, working more or less independently, is much more likely to piece together the truth than is a central, if more "efficient," directorate. They can cite the numerous historical instances in which the greatest discoveries were totally unexpected and unpredictable and thus would have eluded planned research. Thus, discussion of limitations on the resources to be made available to science elicits alarm but little intellectual interest and no sympathy. And discussion about the allocation of finite resources is met with avoidance responses.

But when the cost of basic scientific research is measured in tens of billions of dollars, other social institutions are certain to raise questions. For society, basic scientific research is, like education, an investment in the future, a diversion of resources in money and minds that could otherwise be used to meet present needs. Therefore, society is entitled to ask: What are the prospects, however tenuous or remote, of future benefit? On what time scale and of what magnitude are these benefits, however large the standard degree of error in prediction? And to what extent are such factors considered in the allocation of research resources? Because of apathy, or even antipathy, toward such questions in the scientific community, the answer to the last question would have to be "precious little." Broad resource allocations are based on historical precedent, on political pressure by interest groups (e.g., AIDS activists and pro-agriculture groups), on indigenous cost elements such as expensive instrumentation, and on general public appeal or concern over issues like health, defense, and space exploration.

Is this problem, once confronted, really so intractable? Can weighting factors be devised to be applied, not within disciplines, but over broad domains of science to help design a more rational means of allocation? Implicitly or explicitly, such intellectual discipline is of necessity applied in other costly areas of national importance such as defense, health care, and education. Science can hardly expect to be exempt. And the continued possibility of an engrossing life in science will depend on finding solutions to this allocation problem and to the much greater problem of developing widespread scientific literacy, the only secure basis for informed support of science.

When humans first emerged on the earth out of the shadowed millennia of the preconscious, they were intellectually newborns. They had no idea of their origins or their potential. Their internal and external worlds were simply given to them without explanation. All was mystery—birth, death, disease, famine, light, dark, the stars, and the tides—all occurred as if staged by unseen hands.

Science has provided the key to explain with ever deeper, more general, and more complete concepts the origins and workings of the planet, of the cosmos, of ourselves and all the other creatures of earth. The biologist, peering ever deeper into the machinery of life, is caught between a growing wonder at its beautiful complexity and adroit ingenuity and a growing conviction that we too are but remarkable machines—a most extraordinary, perhaps unparalleled creation, yet a transient assortment of atoms playing a role in a drama whose point we can never fathom.

Just as living is a daily denial of the ultimate reality of death, an absorption in the accessible events of daily life while avoiding the intractable issue of mortality, so is science a denial of the ultimate questions of human or cosmic purpose by absorption in the accessible mysteries of the natural world around and within us.

～～～～～

Why does it seem so strange to have reached an age when one can no longer view death with the detachment of youth? In January 1990, for the seventh time I changed the decade as I wrote the date. How could another decade have slipped by so quickly? Will there be another?

To the students in my classes, World War II is ancient history, quite as remote as the Civil and Revolutionary Wars, if not those between Rome and Carthage and Athens and Troy. The living memory of that cataclysm is fast dying and yet, beyond their ken, that vast conflict shaped my world and theirs as well. From the times of the Greeks and Romans, the elders have looked askance at the mores of the young. Knowing this, I nevertheless feel out of place in an age of lotteries, rock stars, and television violence. Was it better in my youth, with bootleggers and the Teapot Dome and Hollywood scandals? Somehow, I think so. The mindless spectacle, tawdriness, and criminality were much less pervasive and much further from general acceptance.

I feel out of place in a society in which self-actualization support groups and telephone hotlines have replaced the senses of duty, self-reliance, and personal responsibility, in which "society" and "the system" are always the culprit, as if the individual had no free will. Human actions derive from some mixture of genetic determinants, social determinants, and volition. The middle set currently receives much attention. And of the first, we will in the near future know much more. But, it is the third, the capacity for voluntary action, that makes us human and most distinguishes us from all other species. Human society has been built, and properly so, on the concept of personal responsibility, and the more our research clearly defines the roles and strengths of the biological and social determinants, the more we must develop an internal recognition of volition with its possibilities of choice and associated responsibilities.

In the spring, as graduation day approaches at MIT, there will be a time of alumni reunions, and the fifty-year class will be a center of attention. When I was a student, this knot of elders seemed truly ancient—they had graduated in the 1880s, the remote past. Two years ago, I attended the fiftieth reunion of my own class, the fossils of '41. No doubt, we seemed just as ancient to the students of today—our era just as remote, our hopes, dreams, and concerns just as antiquated, however much they still pulse in our memories.

The cycle of life tempers hopes and dreams with harsh reality. Some prove too remote, some illusory. Some come to fruition tinted with a far less roseate hue. I think of television, of how, when I was a student, we dreamed it would bring enlightenment, education, the great wonders of nature, and the great human achievements to all the world. And how, with seeming inevitability, television has been trivialized and suborned by the truly dominant forces of commerce and politics, the sources of power in a secular age. Could it have been otherwise in our society? Fortunately, new generations arise with their new dreams. For, as reality tempers dreams, so in a crucial counterpoint do dreams temper reality.

~~~~~~~~

I sit here in my study, surrounded by the ghosts of science past. These are not ethereal ghosts, undulating gently and silently in the soft air. No, they are solid and weighty tomes—textbooks, journals, conference proceedings, reviews, reports, theses—arrayed on shelves, stacked in piles, and filed in cabinets. They are the mute but viable record of fifty years of biology, of once-lively discussion, of questions now mostly resolved or identified as mirages, of brilliant concepts dashed on reality and major discoveries unexpectedly stumbled across. Of advances in knowledge year by year by year. They are the record of a golden age in biology, when the search for understanding of living processes finally reached the basal genetic level and began to find the ultimate evolutionary explanations.

Now, in this field little is read or is worth reading that is more than five years old. And so my own articles now sit with these others, unattended, buried ever deeper by the unending flow of new literature. But the old records speak to me yet, for I can hear the voices of the seminar speakers, I still recall the late-night discussions, I still feel the excitements, the disappointments, and the wonder as the tale continually unfolded and the mysteries continually receded into the newer unknown—as they do today.

~~~~~~~~

Some twenty years ago, I visited Sinsheim, the dwelling-place of my ancestors. In the early 1970s, it was still a small German village of perhaps four to five thousand persons some fifty kilometers southeast of Heidelberg. It had been spared much of the destruction of World War

II; the seventeenth-century town hall, the old church, and many medieval buildings and dwellings were still in use. Nestled along the tree-shaded Elbenz River, Sinsheim is located in a rolling countryside of small farms and orchards. On a hill some two or three kilometers away are the ruins of an old castle-fort that once protected the inhabitants. My ancestors lived here, for how many centuries? They must have farmed the soil or tended the orchards. Their world was small and bounded. The nearest town, Heidelberg, was two days' journey.

I, their descendant, have lived a very different life, in laboratories and classrooms, part of the worldwide community of science. As part of that community my travels have taken me to most of the continents of earth. I have had so many more opportunities than they. I have been permitted a life of investigation, a life of investment in the future through science and education. I expect they too invested in the future, to the extent then possible, and I know with gratitude that their investment made possible my life and my contributions.

〜〜〜〜〜

Writing this book has been a pilgrimage to faces and scenes of long ago, to dates once future and now past. To seek order in a life, to find meaning in a trajectory. Looked at as the life-span of a scientist-scholar, there are two abrupt shifts or zigzags in the arc. One, to the Radiation Laboratory, was imposed on me and reversed when feasible. The other, to be chancellor at UC Santa Cruz, I chose out of a naïve if lofty idealism. It was too late in life to ever fully reverse.

I find it as yet difficult to assess the decade at Santa Cruz. I learned so much—perhaps that is its own reward. I tried to use all that I learned and all that I knew before to guide the campus toward its true potential. Was my role good for the institution? I firmly believe it was. Was it good for the students and the future of education? I like to believe so, but only history will tell. In that decade, Santa Cruz sought to educate in some degree between ten and fifteen thousand students. How can one measure that impact?

Was it worthwhile for me personally? I have to acknowledge that, for many reasons, much that I had hoped to do was not accomplished. Could I have accomplished more, for my personal satisfaction or for the long-run benefit of humanity, had I remained within science? Who can say? Surely, I became wiser in the ways of the world. Perhaps that is enough.

Both deviations gave me broader experiences, which I treasure, but both I believe diminished my overall scientific productivity—what I might have accomplished in science. Science has given my life a continuity and a thrust. Whatever disillusion has accompanied its triumphs has not dimmed my appreciation of its beauty and power—it has only deepened my compassion for the plight of our species which can both conceive it and so misuse it. Some deep impulse of social concern, some underlying bent of conscience has repeatedly nudged me out of the happy cloistered world of science. To add to knowledge was not enough. Knowledge is not inert; it is a seed that will be used for good or ill. A scientist cannot and should not control that use, but to ignore or pretend ignorance of its potential is to abdicate a central part of one's humanity.

〰〰〰〰

We have only scratched the surface. The great mysteries remain. The mysteries of the cosmos, of the origin of the universe, of "dark matter," of quasars and all the quandaries of galaxies. The mysteries of matter: What lies beyond the quarks? What determines the mass of the proton? The mysteries of evolution, of the origin of life and the origin of man. The mysteries of the mind, of the origin of consciousness and the neurobiology of thought and sensation. They beckon the human spirit and they summon us to endless adventure, to the "endless frontier."

What we have in my lifetime learned about genetics, and what we are today learning about our own human inheritance, will endure. It will endure as long as humans ask questions, as long as we seek some control over our destiny, as long as we know wonder. I have no desire for immortality, but I would love to return in a century or two to see where science stands and to learn what questions they are asking in the Sinsheimer Laboratory.

Glossary

Absorption spectrum	The relative strength of light absorption by a substance as a function of the wavelength of the radiation.
Amino acids	The molecular subunits of proteins.
Amino : imino	A tautomeric equilibrium between amino ($-NH_2$) and imino ($=NH$) forms.
Antigen	A substance that reacts with antibodies or cells of the immune system.
Atomic force microscope	A microscope that scans the topography of surfaces on a near-atomic scale with an extremely sensitive probe.
Bacteriophage	Viruses that attack bacterial cells. T2, T4, and ϕX are specific varieties of bacteriophage.
Biophysics	The science involving the application of physical principles and methods to the study of the structures and processes of living organisms.
Cellular ultrastructure	Assemblies of varied submicroscopic filaments forming scaffolds and shaping substructure in cells of higher organisms, called ultrastructure.
Chloroplast	A small intracellular body containing chlorophyll that plays an essential role in photosynthesis.
Chromosomes	Threadlike structures, each of which carries a linear array of genes.
Clone	A set of genetically identical individual organisms. In man, identical twins are a clone of two.

Dalton	A unit of mass appropriate for individual atoms or molecules, defined as one twelfth of the mass of an atom of the carbon isotope 12; equivalent to 1.66×10^{-24} grams.
Density gradient centrifugation	A technique of centrifugation in which the centrifugal field is employed to establish a gradient of density in a solvent; particles of different density then separate in the centrifuge, each settling in the gradient at a point equal to its own density.
Density label	A means of labeling one group of molecules by altering its density relative to another similar group; often used in conjunction with density gradient centrifugation.
Deoxyribonucleic acids	*See* Nucleic acids.
Developmental biology	That part of biology which concerns the progression of an organism from the fertilized egg to the adult form.
DNA ligase	An enzyme that can link together DNA chains.
DNA	*See* Nucleic acids.
DNA polymerase	An enzyme that catalyzes the linking of nucleotides into a DNA chain.
DNA renaturation	The process by which two strands of a DNA double helix, having been separated in solution, selectively rejoin each other and pair up over time.
DNA repair	Mechanisms to effect the repair of chemical or mechanical damage to a cell's DNA; all cells have a variety of such processes.
Drosophila	A specific family of fruit flies used extensively in genetic research.
Electromagnetic radiation	Radiation composed of electromagnetic waves or, alternatively, photons; the electromagnetic spectrum comprises the whole range of electromagnetic wavelengths or photon energies.
Enzyme	A protein that serves as a catalyst for a particular biochemical reaction.
Eukaryote	A cell with a definitive nucleus and mitochondria.
Exon : intron	Alternating tracts comprising the messenger RNA molecules in higher organisms as initially transcribed from DNA. The introns are subsequently deleted and the exons spliced together to form the functional messenger RNA molecules.
Gene	The functional unit of inheritance.
Genetic code	The code linking nucleotide sequences in nucleic acids with amino acid sequences in proteins.
Genetic map	A graphic presentation of the linear order and spacing of genes in a chromosome.

Genome	The complete genetic endowment of a member of a species.
Heavy water	Water in which all of the normal hydrogen atoms (atoms of the hydrogen isotope 1) have been replaced by deuterium atoms (atoms of the hydrogen isotope 2).
Infrared spectroscopy	*See* Spectroscopy.
Ionizing radiation	Radiation composed of particles or photons of sufficient energy to remove electrons from the atoms or molecules of the material through which they pass, leaving a trail of charged atoms or molecules, that is, a trail of ions.
Isomer	One of two or more chemical substances having the same atomic composition but differing in structure.
Isotope	One of two or more atoms of an element, differing in mass. Nonradioactive isotopes are stable. Radioactive isotopes are unstable and decay by radioactivity.
Keto : enol	A tautomeric equilibrium between keto ($=O$) and enol ($-OH$) forms.
Light scattering	The absorption and instantaneous reemission in varied directions of light without change of wavelength.
Messenger RNA	An RNA molecule the nucleotide sequence of which parallels that of one strand of a portion of a double-helical DNA molecule and that programs protein synthesis on a ribosome.
Microdensitometer	An instrument to measure and compare the transmission of light by microscopic areas of a specimen.
Micron	A unit of length convenient for cellular dimensions equal to one millionth of a meter.
Microwaves	Electromagnetic radiation with wavelengths generally between 0.3 and 30 centimeters.
Mitochondria	Small bodies found in all eukaryotic cells that play an essential role in the provision of energy.
Mitosis	Division of a cellular nucleus into two with exact duplication and separation of the chromosomes.
Molecular biology	That part of biology which attempts to interpret biological events in terms of the physico-chemical properties of the molecules (especially the macromolecules) in a cell.
Nanometer	A unit of length appropriate to molecular or crystalline dimensions; equal to one billionth of a meter.

Nucleic acids	Large linear molecules composed of nucleotide subunits joined by phosphate linkages between the sugar moieties. A string of nucleotides is called an *oligonucleotide* or *polynucleotide*. There are two general types:

In *deoxyribonucleic acids* (DNA), the sugar moiety of the nucleotide subunits is always deoxyribose. DNA molecules are composed of four principal nucleotide subunits: deoxyadenylic acid, deoxyguanylic acid, deoxycytidylic acid, and thymidylic acid. In some DNAs, a small amount of a fifth nucleotide, 5-methyldeoxycytidylic acid, replaces part of the deoxycytidylic acid.

In cells, DNA is invariably present as a double-helical structure, composed of two intertwined polynucleotide strands, in a paired fashion. DNA in some viruses is single-stranded.

Genes are composed of DNA, except in certain viruses which use RNA.

In *ribonucleic acid* (RNA), the sugar moiety of the nucleotide subunits is always ribose. RNA molecules are composed of four varieties of nucleotide subunits: adenylic acid, guanylic acid, cytidylic acid, and uridylic acid.

In cells, RNA is always single-stranded. Double-stranded RNA is found only in certain viruses and RNA virus infections.

Nucleotide	A subunit of nucleic acid. Each nucleotide is composed of an organic ring-shaped molecule (either a pyrimidine-type ring or a purine-type ring) attached to a five-carbon sugar attached in turn to a phosphate group.
Oligonucleotide	*See* Nucleic acids.
Photoreactivation	A phenomenon in which some of the effects of ultraviolet radiation on cells can subsequently be reversed by exposure to visible light.
Plasmid	A small extrachromosomal genetic element in cells.
Polynucleotide	*See* Nucleic acids.
Proteins	High molecular weight molecules composed of amino acids joined primarily by peptide linkages.
Pulsar	A celestial radio source emitting short bursts of radio emission at regular intervals.
Purine	A nine-membered double ring molecule composed of carbon and nitrogen atoms with varied side groups; the purines found in nucleic acids are adenine and guanine.

Pyrimidine	A six-membered ring composed of carbon and nitrogen atoms with varied side groups; the pyrimidines found in nucleic acids are cytosine, 5-methylcytosine, uracil, and thymine.
Quark	A basic particle of physics with a charge of one or two thirds of the electron charge; presumed component of nucleons.
Quasar	A quasi-stellar astronomical object of great brilliance and radio intensity, often with large redshift.
Raman spectroscopy	*See* Spectroscopy.
Recombinant DNA	DNA produced by splicing together DNA molecules of different origin.
Recombinant, genetic	An organism in which genes from two sources have been recombined into one chromosome.
Refractive index	The ratio of the velocity of light in a vacuum to that in a medium; the difference in the refractive indices of two media is determinant of the amount of light reflected at their interface.
Restriction enzyme	An enzyme that can cleave DNA at a specific nucleotide sequence.
Ribonucleic acid	*See* Nucleic acids.
Ribosomal RNA	RNA molecules that form an essential part of the structure of the ribosomes.
Ribosomes	Small complex particles, found in all cells, on which protein synthesis takes place.
RNA	*See* Nucleic acids.
Saturated hydrocarbon	A compound composed of carbon and hydrogen atoms with all carbon bonds filled.
Semiconservative replication	A term applied to replication of double-stranded DNA molecules, in which the two strands of the parental DNA remain intact but are separated, each passing to one of the two daughter DNA molecules.
Spectroscopy	The science of the production, measurement, and interpretation of electromagnetic radiation as a function of its wavelength. *Infrared spectroscopy* is concerned with radiation in the infrared region, typically between 0.7 and 25 microns in wavelength; *ultraviolet spectroscopy* is concerned with radiation in the ultraviolet region, typically between 0.2 and 0.4 microns in wavelength; *Raman spectroscopy* is concerned with radiation reemitted from a transparent medium with changed wavelength, indicative of molecular energy levels and structure.

Transcription	The process of forming a messenger RNA molecule that conveys the information content of a portion of a DNA molecule.
Transducer	A device that converts an input signal into an output signal of a different form, for example, a microphone.
Transfer RNA	Small RNA molecules that play an essential role in the translation via the genetic code from nucleotide sequence in nucleic acid to amino acid sequence in protein.
Transforming principle	A substance that effects specific genetic transformation in bacterial cells—known to be DNA.
Ultracentrifuge	A high-speed centrifuge capable of developing centrifugal force in excess of 100,000 times gravity.
Ultraviolet spectroscopy	*See* Spectroscopy.
X-ray diffraction	The scattering of X-rays by crystals with resultant interference effects dependent on the spacing and atomic composition of the components of the crystals.

Selected Personal References

World War II—The Radiation Laboratory

"Altitude determination," in *Radar Aids to Navigation,* ed. J. S. Hall (Mc-Graw-Hill, 1947), 131–42.
"Design of a lightweight airborne radar for navigation," in *Radar System Engineering,* ed. L. N. Ridenour (McGraw-Hill, 1947), 616–25.
"Lightweight airborne receiver," in *Radar System Engineering,* ed. L. N. Ridenour (McGraw-Hill, 1947), 464–70.
"Performance of AN/APS-10," in *Radar Aids to Navigation,* ed. J. S. Hall (McGraw-Hill, 1947), 171–85.

Graduate School

"Christiansen filters for the ultra-violet," R. L. Sinsheimer and J. R. Loofbourow, *Nature* 160 (1947): 674–75.
"Low temperature spectroscopy of biological compounds," J. F. Scott, R. L. Sinsheimer, and J. R. Loofbourow, *Science* 107 (1948): 302.
"Use of thin films of sublimate for absorption spectroscopy," R. L. Sinsheimer, J. F. Scott, and J. R. Loofbourow, *Nature* 164 (1949): 796–97.
"Microscopy: Ultraviolet and microabsorption spectroscopy," J. F. Scott and R. L. Sinsheimer, *Medical Physics,* ed. O. Glasser (Yearbook Publishers, Inc., 1950), 2:537–50.
"Ultraviolet absorption spectra at reduced temperatures: Principles and meth-

ods," R. L. Sinsheimer, J. F. Scott, and J. R. Loofbourow, *Journal Biological Chemistry* 187 (1950): 299–312.

"Ultraviolet absorption spectra at reduced temperatures: Pyrimidines and purines," R. L. Sinsheimer, J. F. Scott, and J. R. Loofbourow, *Journal Biological Chemistry* 187 (1950): 313–24.

"Factors involved in the sharpening of the ultraviolet absorption spectrum of guanine at reduced temperature," J. F. Scott, R. L. Sinsheimer, and J. R. Loofbourow, *Journal American Chemical Society* 74 (1952): 275–77.

Iowa State—Research and Discovery

"Ion exchange separation of desoxyribonucleotides," R. L. Sinsheimer and J. F. Koerner, *Science* 114 (1951): 42–43.

"A purification of venom phosphodiesterase," R. L. Sinsheimer and J. F. Koerner, *Journal Biological Chemistry* 198 (1952): 293–96.

"Light scattering by tobacco mosaic virus nucleic acid," T. G. Northrop and R. L. Sinsheimer, *Journal of Chemical Physics* 22 (1954): 703–7.

"Infrared absorption spectra of pyrimidine nucleotides in H_2O and D_2O solution," R. L. Sinsheimer, R. L. Nutter, and G. R. Hopkins, *Biochimica & Biophysica Acta* 18 (1955): 13–27.

A Sidelight on Watson and Crick

"Di-desoxyribonucleotides," R. L. Sinsheimer and J. F. Koerner, *Journal American Chemical Society* 74 (1952): 283.

"The action of pancreatic deoxyribonuclease: Isolation of mono- and dinucleotides," R. L. Sinsheimer, *Journal Biological Chemistry* 208 (1954): 445–59.

"The action of pancreatic deoxyribonuclease: Isomeric dinucleotides," R. L. Sinsheimer, *Journal Biological Chemistry* 215 (1955): 579–83.

Iowa State—At Full Speed

"Nucleotides from T2r+ bacteriophage," R. L. Sinsheimer, *Science* 120 (1954): 551–53.

"The photochemistry of uridylic acid," R. L. Sinsheimer, *Radiation Research* 1 (1954): 505–13.

"Technique of study of biological effects of ultraviolet radiation," J. F. Scott and R. L. Sinsheimer, in *Radiation Biology*, ed. A. Hollaender (McGraw-Hill, 1955), vol. 2, ch. 4:119–63.

"Ultraviolet absorption spectra," R. L. Sinsheimer, in *Radiation Biology*, ed. A. Hollaender (McGraw-Hill, 1955), vol. 2, ch. 5:165–201.

"Visible and ultraviolet light scattering by tobacco mosaic virus nucleic acid," G. R. Hopkins and R. L. Sinsheimer, *Biochimica & Biophysica Acta* 17 (1955): 476–84.

"The glucose content of the deoxyribonucleic acids of certain bacteriophages," R. L. Sinsheimer, *Proceedings of National Academy of Sciences* 42 (1956): 502–4.

"A deoxyribonuclease from calf spleen: Purification and properties," J. F. Koerner and R. L. Sinsheimer, *Journal Biological Chemistry* 228 (1957): 1039–48.

"A deoxyribonuclease from calf spleen: Mode of Action," J. F. Koerner and R. L. Sinsheimer, *Journal Biological Chemistry* 228 (1957): 1049–62.

"First steps toward a genetic chemistry," R. L. Sinsheimer, *Science* 125 (1957): 1123–28.

"The photochemistry of cytidylic acid," R. L. Sinsheimer, *Radiation Research* 6 (1957): 121–25.

"Studies upon DNA synthesis during multiplicity reactivation of T2r+ bacteriophage," R. L. Nutter and R. L. Sinsheimer, *Virology* 7 (1959): 276–90.

ϕX

ϕX STRUCTURE

"Purification and properties of bacteriophage ϕX174," R. L. Sinsheimer, *Journal Molecular Biology* 1 (1959): 37–42.

"A single-stranded deoxyribonucleic acid from bacteriophage ϕX174," R. L. Sinsheimer, *Journal Molecular Biology* 1 (1959): 43–53.

"The structure of the DNA of bacteriophage ϕX174: III. Ultracentrifugal evidence for a ring structure," W. Fiers and R. L. Sinsheimer, *Journal Molecular Biology* 5 (1962): 424–34.

"Electron microscopy of the replicative form of the DNA of the bacteriophage ϕX174," A. K. Kleinschmidt, A. Burton, and R. L. Sinsheimer, *Science* 142 (1963): 961.

"The structure of the DNA of bacteriophage ϕX174: IV. Pyrimidine sequences," J. B. Hall and R. L. Sinsheimer, *Journal Molecular Biology* 6 (1963): 115–27.

"The process of infection with bacteriophage ϕX174: XXVIII. Removal of the spike proteins from the phage capsid," M. H. Edgell, C. A. Hutchison III, and R. L. Sinsheimer, *Journal Molecular Biology* 42 (1969): 547–57.

"A cleavage map of bacteriophage ϕX174 genome," A. S. Lee and R. L.

Sinsheimer, *Proceedings of National Academy of Sciences* 71 (1974): 2882–86.

"The location of the 5-methylcytosine group on the bacteriophage φX174 genome," A. S. Lee and R. L. Sinsheimer, *Journal of Virology* 14 (1974): 872–77.

"Aligning the φX174 genetic map and the φX174 heteroduplex denaturation map," J. L. Compton and R. L. Sinsheimer, *Journal Molecular Biology* 109 (1977): 217–34.

φX REPLICATION

"Infection of protoplasts of *Escherichia coli* by subviral particles of bacteriophage φX174," G. D. Guthrie and R. L. Sinsheimer, *Journal Molecular Biology* 2 (1960): 297–305.

"The process of infection with bacteriophage φX174: I. Evidence for a 'replicative form,' " R. L. Sinsheimer, B. Starman, C. Nagler, and S. Guthrie, *Journal Molecular Biology* 4 (1962): 142–60.

"Kinetics of bacteriophage release by single cells of φX174-infected *E. coli*," C. A. Hutchison and R. L. Sinsheimer, *Journal Molecular Biology* 7 (1963): 206–8.

"The process of infection with bacteriophage φX174: VII. Ultracentrifugal analysis of the replicative form," A. Burton and R. L. Sinsheimer, *Journal Molecular Biology* 14 (1965): 327–47.

"Enzymatic synthesis of DNA: XXIV. Synthesis of infectious phage φX174 DNA," M. Goulian, A. Kornberg, and R. L. Sinsheimer, *Proceedings of National Academy of Sciences* 58 (1967): 2321–28.

"The process of infection with bacteriophage φX174: XVI. Synthesis of the replicative form and its relationship to viral single-stranded DNA synthesis," B. H. Lindqvist and R. L. Sinsheimer, *Journal Molecular Biology* 32 (1968): 285–302.

"The process of an infection with bacteriophage φX174: XXI. Replication and fate of the replicative form," R. Knippers, T. Komano, and R. L. Sinsheimer, *Proceedings of National Academy of Sciences* 59 (1968): 577–81.

"The process of infection with bacteriophage φX174: XXII. Synthesis of progeny single-stranded DNA," T. Komano, R. Knippers, and R. L. Sinsheimer, *Proceedings of National Academy of Sciences* 59 (1968): 911–16.

"The process of infection with bacteriophage φX174: XXIX. In vivo studies on the synthesis of the single-stranded DNA of progeny φX174 bacteriophage," R. Knippers, A. Razin, R. Davis, and R. L. Sinsheimer, *Journal Molecular Biology* 45 (1969): 237–63.

"Bacteriophage φX174 DNA synthesis in a replication-deficient host: Determination of the origin of φX174 DNA replication," P. D. Baas, H. S. Jansz, and R. L. Sinsheimer, *Journal Molecular Biology* 102 (1976): 633–56.

φX Transcription

"The in vivo φX mRNA," J. W. Sedat and R. L. Sinsheimer, *Cold Spring Harbor Symposia on Quantitative Biology* 35 (1970): 163–70.

"Nucleotide sequences of the 5′ termini of φX174 mRNAs synthesized in vitro," L. H. Smith, K. Grohmann, and R. L. Sinsheimer, *Nucleic Acids Research* 1 (1974): 1521–29.

"The in vitro transcription units of φX174: Initiation with specific 5′ end oligonucleotides of in vitro φX174 RNA," L. H. Smith and R. L. Sinsheimer, *Journal Molecular Biology* 103 (1976): 711–35.

φX Genetics

"The process of infection with bacteriophage φX174: X. Mutants in a φX lysis gene," C. A. Hutchison III and R. L. Sinsheimer, *Journal Molecular Biology* 18 (1966): 429–47.

"Genetic expression in heterozygous replicative form molecules of φX174," V. Merriam, L. Dumas, and R. L. Sinsheimer, *Journal Virology* 7 (1971): 603–11.

"The genetic map of bacteriophage φX174," R. M. Benbow, C. A. Hutchison III, J. D. Fabricant, and R. L. Sinsheimer, *Journal Virology* 7 (1971): 549–58.

"Direction of translation and size of bacteriophage φX174 cistrons," R. M. Benbow, R. F. Mayol, J. C. Picchi, and R. L. Sinsheimer, *Journal Virology* 10 (1972): 99–114.

"Recombinant DNA molecules of bacteriophage φX174," R. M. Benbow, A. J. Zuccarelli, and R. L. Sinsheimer, *Proceedings of National Academy of Sciences* 72 (1975): 235–39.

φX Summaries

"Single-stranded DNA," R. L. Sinsheimer, *Scientific American* (July 1962): 109–16.

"Rings of life—and death," R. L. Sinsheimer, *Engineering & Science* (November 1963): 18–20.

"φX174 DNA," R. L. Sinsheimer, in *Procedures in Nucleic Acid Research,* ed. G. L. Cantoni and D. R. Davies (New York: Harper and Row, 1966), 569–76.

"φX: Multum in parvo," R. L. Sinsheimer, in *Phage and Origins of Molecular Biology,* ed. J. Cairns (Cold Spring Harbor Laboratory of Quantitative Biology, 1966), 258–64.

"Bacteriophage φX174: Viral functions," R. L. Sinsheimer, C. A. Hutchison III, and B. Lindqvist, in *The Molecular Biology of Viruses,* ed. L. V. Crawford and M. G. P. Stoker (Academic Press, 1967), 175–92.

"Closing the ring," R. L. Sinsheimer, *Technology Review* (MIT, July/August 1968): 22–27.

"Spheroplast assay of φX174 DNA," R. L. Sinsheimer, in *Methods in Enzymology*, ed. L. Grossman and K. Moldave (New York: Academic Press, 1968), vol. 12, pt. B, 850–58.

"The life cycle of a single-stranded DNA virus (φX174)," R. L. Sinsheimer, *Harvey Lecture Series* 64 (1968–69): 69–86.

"Contagium vivum fluidum, after seventy years," R. L. Sinsheimer (Proceedings of Koninklijke Nederlandse Akademie van Wetenschappen, 1970), ser. C, no. 73, 307–16.

"Bacteriophage φX174," R. L. Sinsheimer, in *Handbook of Genetics*, ed. R. C. King (Plenum Press, 1974): 323–25.

"φX—Some recollections," R. L. Sinsheimer, in *The Single-Stranded Phages* (Cold Spring Harbor Laboratory, 1978), 3–6.

"φX174, a research odyssey: From plaque to particle, and mutant to molecule," R. L. Sinsheimer, in *Genes, Cells and Behavior*, ed. N. H. Horowitz and E. Hutchings, Jr. (San Francisco: W. E. Freeman and Co., 1980), 44–60.

"The Discovery of a Single-Stranded, Circular DNA Genome," *Bioessays* 13, no. 2 (February 1991), 89–91.

The Caltech Years—The Fulfillment

BACTERIOPHAGE MS2

"Purification and properties of bacteriophage MS2 and of its ribonucleic acid," J. H. Strauss and R. L. Sinsheimer, *Journal Molecular Biology* 7 (1963): 43–54.

"The replication of bacteriophage MS2: I. Transfer of parental nucleic acid to progeny phage," J. F. Davis and R. L. Sinsheimer, *Journal Molecular Biology* 6 (1963): 203–7.

"The replication of bacteriophage MS2: IV. RNA components specifically associated with infection," R. B. Kelly, J. L. Gould, and R. L. Sinsheimer, *Journal Molecular Biology* 11 (1965): 562–75.

"Characterization of an infectivity assay for the RNA of bacteriophage MS2," J. A. Strauss, Jr., and R. L. Sinsheimer, *Journal Virology* 1 (1963): 711–16.

"The replication of bacteriophage MS2: VI. Interaction between bacteriophage RNA and cellular components in MS2-infected *Escherichia coli*," G. N. Godson and R. L. Sinsheimer, *Journal Molecular Biology* 23 (1967): 495–521.

"The replication of bacteriophage MS2: VII. Nonconservative replication of double-stranded RNA," R. B. Kelly and R. L. Sinsheimer, *Journal Molecular Biology* 26 (1967): 169–79.

"The replication of bacteriophage MS2: IX. The structure and replication of the replicative intermediate," R. B. Kelly and R. L. Sinsheimer, *Journal Molecular Biology* 29 (1967): 237–49.

"Initial kinetics of degradation of MS2 ribonucleic acid by ribonuclease, heat and alkali and the presence of configurational restraints in this ribonucleic acid," J. Strauss, Jr., and R. L. Sinsheimer, *Journal Molecular Biology* 34 (1968): 453–65.

TOBACCO MOSAIC VIRUS

"Biophysical studies of infectious ribonucleic acid from tobacco mosaic virus," P. C. Cheo, B. S. Friesen, and R. L. Sinsheimer, *Proceedings of National Academy of Sciences* 45 (1959): 305–13.

"Partition cell analysis of infective tobacco mosaic virus nucleic acid," B. S. Friesen and R. L. Sinsheimer, *Journal Molecular Biology* 1 (1959): 321–28.

BACTERIOPHAGE λ

"Novel intra-cellular forms of λ DNA," E. T. Young II and R. L. Sinsheimer, *Journal Molecular Biology* 10 (1964): 562–64.

"Vegetative bacteriophage λ DNA: I. Infectivity in a spheroplast assay," E. T. Young II and R. L. Sinsheimer, *Journal Molecular Biology* 30 (1967): 147–64.

"Vegetative bacteriophage λ DNA: II. Physical characterization and replication," E. T. Young II and R. L. Sinsheimer, *Journal Molecular Biology* 30 (1967): 165–200.

"Purification and properties of intracellular λ DNA rings," J. A. Kiger, Jr., E. T. Young II, and R. L. Sinsheimer, *Journal Molecular Biology* 33 (1968): 395–413.

"DNA of vegetative bacteriophage λ: VI. Electron microscopic studies of replicating Lambda DNA," J. A. Kiger, Jr., and R. L. Sinsheimer, *Proceedings of National Academy of Sciences* 68 (1971): 112–15.

BACTERIOPHAGE PM2

"DNA of bacteriophage PM2: A closed circular double-stranded molecule," R. T. Espejo, E. S. Canelo, and R. L. Sinsheimer, *Proceedings of National Academy of Sciences* 63 (1969): 1164–68.

"A difference between intracellular and viral supercoiled PM2 DNA," R. Espejo, E. Espejo-Canelo, and R. L. Sinsheimer, *Journal Molecular Biology* 56 (1971): 623–26.

REVIEW ARTICLES

"Current ideas of the structure of deoxyribonucleic acid," R. L. Sinsheimer, in *A Symposium on Molecular Biology,* ed. R. E. Zirkle (University of Chicago Press, 1959), ch. 2:16–30.

"The biochemistry of genetic factors," R. L. Sinsheimer, *Annual Review Biochemistry* 29 (1960): 503–24.

"Nucleic acids of the bacterial viruses," R. L. Sinsheimer, in *Nucleic Acids,* ed. E. Chargaff and J. M. Davidson (Academic Press, 1960), 3:187–244.

"The structure of DNA and RNA," R. L. Sinsheimer, in *The Molecular Control of Cellular Activity,* ed. J. M. Allen (McGraw-Hill, 1962), 8:221–43.

"Replication of the nucleic acids of the bacterial viruses," R. L. Sinsheimer, in *Viruses, Nucleic Acids, and Cancer* (Baltimore: Williams and Wilkins Company, 1963), 246–51.

"The replication of viral DNA," R. L. Sinsheimer, in *The Molecular Biology of Viruses,* Society for General Microbiology (Cambridge University Press, 1968), 18:101–23.

"DNA virus maturation," R. L. Sinsheimer, *Journal Cellular Physiology* 74, supp. 1, pt. 2 (1969): 21–32.

"Recombinant DNA," R. L. Sinsheimer, *Annual Review Biochemistry* 46 (1977): 415–38.

Science and Society—Toward Wider Horizons

THE FUTURE OF BIOLOGY AND MEDICINE

The Book of Life, R. L. Sinsheimer (Reading, Mass.: Addison-Wesley Publishing Co., 1967).

"The end of the beginning," R. L. Sinsheimer, *Bulletin of Atomic Scientists,* vol. 23, no. 2 (February 1967): 8–12.

"Darkly wise and rudely great," R. L. Sinsheimer, *Engineering & Science* (CIT, May 1968), 20–28.

"The prospect for designed genetic change," R. L. Sinsheimer, *Engineering & Science* (April 1969); reprinted in *American Scientist* 57 (1969): 134–42.

"Genetic engineering: The modification of man," R. L. Sinsheimer, *Impact of Science on Society* 20 (1970): 279–89.

"The implications of recent advances in biology for the future of medicine," R. L. Sinsheimer, *Engineering & Science* (October 1970).

"Whither Molecular Biology," R. L. Sinsheimer, *Engineering & Science* (March/April 1977), 29–34.

"Genetic engineering and gene therapy: Some implications," R. L. Sinsheimer, in *Genetic Issues in Public Health and Medicine,* ed. B. H. Cohen, A. M. Lilienfeld, and P. C. Huang (Springfield, Ill.: C. C. Thomas, 1978), ch. 21:429–61.

"The awesome powers we face," R. L. Sinsheimer, *American Medical News* (26 January 1979).

"Key Disciplines of the New Biology: Genetics, Immunology, and Neurobiology," R. L. Sinsheimer (delivered at the conference sponsored by the University of North Carolina and the Josiah Macy, Jr., Foundation), in *The New Biology and Medical Education: Merging the Biological, Information, and Cognitive Sciences*, ed. C. P. Friedman and E. F. Powell (Independent Publishers Group, 1983).

Recombinant DNA and Genetic Engineering

"The Dilemma of DNA," R. L. Sinsheimer, in *Wednesday Night at the Lab: Antibiotics, Bioengineering, Contraceptives, Drugs and Ethics*, ed. K. L. Rinehart, Jr., W. O. McClure, and T. L. Brown (Harper and Row, 1973), 121–34.

"Genetic engineering: Ambush or opportunity?" R. L. Sinsheimer, *Queens Quarterly*, vol. 80, no. 2 (1973): 157–79.

"Troubled dawn for genetic engineering," R. L. Sinsheimer, *New Scientist* 68 (1975): 148–51.

"The hazards of recombinant DNA," R. L. Sinsheimer, *Trends in Biochemical Sciences*, vol. 1, no. 8 (August 1976).

"Recombinant DNA—On Our Own," R. L. Sinsheimer, *Bioscience* (October 1976), 599.

"An evolutionary perspective for genetic engineering," R. L. Sinsheimer, *New Scientist* 73 (1977): 150–52.

"Mapping the mammalian genome: Potential risks," *Research with Recombinant DNA*, R. L. Sinsheimer (delivered at the Academy Forum of the National Academy of Sciences, Washington, D.C., 7–9 March 1977), 74–89.

"Prospects and dangers of experimentation with recombinant DNA," R. L. Sinsheimer, *Grenzen der Forschung* (Colloquium Verlag Berlin, 1980), 65–75.

"Genetic engineering: Life as a plaything," R. L. Sinsheimer, *Technology Review*, vol. 86, no. 3 (April 1983): 14–15, 70.

"Recombinant DNA and Biological Warfare," Susan Wright and R. L. Sinsheimer, *Bulletin of the Atomic Scientists* (November 1983), 20–26.

"Reminiscences of the Recombinant DNA Story," with comments by R. L. Sinsheimer, *Bio Essays*, vol. 1, no. 2 (1984): 83–84.

"Biotechnology: The Public Concerns," R. L. Sinsheimer, *Journal Minnesota Academy of Sciences*, vol. 53, no. 1 (1987): 16–20.

Philosophical Implications

"The brain of Pooh: An essay on the limits of mind," R. L. Sinsheimer, *Engineering & Science* (January 1970): 8–13, 36–39; reprinted in *American Scientist* (January/February 1971).

"Science and the Quest for Human Values," R. L. Sinsheimer, in *Science and Human Values in the Twenty-first Century,* ed. Ralph W. Burhoe, (Westminster Press, 1971): 116–34.

"A Time of Metamorphosis," R. L. Sinsheimer, *Engineering & Science* (May/June 1971), 18–20.

"Echoes of the future," R. L. Sinsheimer, *The Journal of the Blaisdell Institute,* vol. 7, nos. 3 and 4 (1972): 4–12.

"The Molecular Basis of Life," R. L. Sinsheimer, in *The Heritage of Copernicus: Theories "More Pleasing to the Mind,"* ed. J. Neyman (MIT Press, 1974), 143–65.

"On Coupling Inquiry and Wisdom," R. L. Sinsheimer (delivered at the 1976 Meeting of the American Society of Biological Chemists, San Francisco, 8 June 1976), *Federation Proceedings* 35 (1976): 2540–42.

"An inquiry into inquiry," R. L. Sinsheimer, *Engineering & Science,* California Institute of Technology (May/June 1976), 15–17; reprinted in *The Hastings Center Report* 6 (1976): 18.

"Humanism and Science," R. L. Sinsheimer, *Engineering & Science* (CIT, October/November 1975), 10–13, reprinted in *Leonardo* 10 (Pergamon Press, 1977), 59–62.

"The Galilean Imperative," R. L. Sinsheimer, in *Recombinant DNA: Science, Ethics, and Politics,* ed. John Richards (Academic Press, 1978), 17–32.

"Genetic Intervention and Values: Are All Men Created Equal?" R. L. Sinsheimer, in *Modifying Man: Implications and Ethics,* ed. Craig W. Ellison (University Press of America, 1978), 109–35.

"The presumptions of science," R. L. Sinsheimer, *Daedalus* (1978): 23–35.

"Technology can free, but it can also impose its own constraints," R. L. Sinsheimer, in *Report on the Douglas Convocation of the State of Individual Freedom* (held 7–8 December 1978), under "Freedom and the New Property," 61–63; reprinted in *Center Magazine* (Santa Barbara, Calif., 1979).

"The Answer Is Not Necessarily the Solution," R. L. Sinsheimer, *Engineering & Science,* CIT, vol. 52, no. 1 (CIT, Fall 1988): 28–31.

"The Responsibility of Scientists," in *Preventing A Biological Arms Race,* ed. Susan Wright (Cambridge, Mass.: MIT Press, 1990), ch. 3.

The Faculty

"New Genetics Industry Tests University Values," R. L. Sinsheimer, *Center Magazine* (May/June 1983), 43–50.

"Peer Review and the Public Interest," commentary by R. L. Sinsheimer, *Issues in Science and Technology,* vol. 2, no. 1 (Fall 1985).

The Telescope and the Genome

"The Santa Cruz Workshop—May, 1985," R. L. Sinsheimer, *Genomics* 5, (1989): 954–56.

"The Potential of Maps," R. L. Sinsheimer, letter to *Technology Review*, vol. 93, no. 4 (May/June 1990): 10.

"Whither the Genome Project?" R. L. Sinsheimer, *Hastings Center Report* 20, no. 4 (July/August 1990): 5.

"The Human Genome Initiative," R. L. Sinsheimer, editorial in *Federation of American Societies for Experimental Biology Journal* 5 (November 1991): 2885.

Return to Science

"Imaging Single-Stranded DNA, Antigen-Antibody Reaction and Polymerized Langmuir-Bladgett Films with an Atomic Force Microscope," A. L. Weisenhorn, H. E. Gaub, H. G. Hansma, R. L. Sinsheimer, G. L. Kelderman, and P. K. Hansma, *Scanning Microscopy* 4 (1990): 511–16.

"Reproducible Imaging and Dissection of Plasmid DNA under Liquid with the Atomic Force Microscope," H. G. Hansma, J. Vesenka, C. Siegerist, G. Kelderman, H. Morrett, R. L. Sinsheimer, V. Elings, C. Bustamente, and P. K. Hansma, *Science* 256 (1992): 1180–84.

Index

Compositor: Impressions, A Division of
            Edwards Brothers, Inc.
   Printer: Edwards Brothers, Inc.
   Binder: Edwards Brothers, Inc.
     Text: 10/13 Galliard
  Display: Galliard